The Role of 5-HT Systems on Memory and Dysfunctional Memory

The Role of 5-HT Systems on Memory and Dysfunctional Memory

Emergent Targets for Memory Formation and Memory Alterations

Alfredo Meneses
Department of Pharmacobiology,
CINVESTAV, Mexico

AMSTERDAM • BOSTON • HEIDELBERG • LONDON
NEW YORK • OXFORD • PARIS • SAN DIEGO
SAN FRANCISCO • SINGAPORE • SYDNEY • TOKYO

ELSEVIER

Academic Press is an imprint of Elsevier

Academic Press is an imprint of Elsevier
32 Jamestown Road, London NW1 7BY, UK
The Boulevard, Langford Lane, Kidlington, Oxford, OX5 1GB, UK
Radarweg 29, PO Box 211, 1000 AE Amsterdam, The Netherlands
225 Wyman Street, Waltham, MA 02451, USA
525 B Street, Suite 1900, San Diego, CA 92101-4495, USA

First published 2014

Notices
Knowledge and best practice in this field are constantly changing. As new research and
experience broaden our understanding, changes in research methods, professional practices,
or medical treatment may become necessary.

Practitioners and researchers must always rely on their own experience and knowledge in
evaluating and using any information, methods, compounds, or experiments described herein.
In using such information or methods they should be mindful of their own safety and the safety
of others, including parties for whom they have a professional responsibility.

To the fullest extent of the law, neither the Publisher nor the authors, contributors, or editors,
assume any liability for any injury and/or damage to persons or property as a matter of products
liability, negligence or otherwise, or from any use or operation of any methods, products,
instructions, or ideas contained in the material herein.

British Library Cataloguing-in-Publication Data
A catalogue record for this book is available from the British Library

Library of Congress Cataloging-in-Publication Data
A catalog record for this book is available from the Library of Congress

ISBN: 978-0-12-800836-2

For information on all Academic Press publications
visit our website at **store.elsevier.com**

This book has been manufactured using Print On Demand technology. Each copy is produced
to order and is limited to black ink. The online version of this book will show color figures
where appropriate.

ELSEVIER   Book Aid
International

**Working together
to grow libraries in
developing countries**

www.elsevier.com • www.bookaid.org

DEDICATION

To my wife Erika, who shows me how wonderful life and love can be. To our daughter Sofia, whom we admire and love. Sofia I know you will become a bright scientist, shining in your field.

To our two little Angels.

CONTENTS

ACKNOWLEDGMENTS

I thank Sofia Meneses-Goytia for revising language and Roberto Gonzalez for his expert assistance. This work was supported in part by CONACYT grant 80060.

CHAPTER 1

Introduction

Drugs acting through 5-hydroxytryptamine (serotonin or 5-HT) systems modulate memory and its alterations, although the mechanisms involved are poorly understood. The neurotransmitter 5-HT was discovered more than 50 years ago, and currently it still continues to generate interest as one of the most successful targets for therapeutic applications (e.g., depression, schizophrenia, anxiety, learning, and memory disorders) (Nordquist and Oreland, 2010; Ruiz and Oranias, 2010). Memory had been classified according to content, time, and its neuroanatomical and biological basis (Meneses, 2013, 2014; Meneses et al., 2011a,b). Diverse brain areas (hippocampus, prefrontal cortex (PFC), etc.) and neurotransmission systems mediate memory systems, including the cholinergic, glutamatergic, dopaminergic, and serotonergic (Burghardt and Bauer, 2013; Cassel, 2010; Eppinger and Hämmerer, 2012; Meneses, 2014; Rodríguez et al., 2012; Singh et al., 2013), and this notion has gained wider acceptance and interest. It is well known that serotonin plays a central role in neural plasticity using different 5-HT receptors (Bockaert et al., 2010; Millan, 2011; Mnie-Filali et al., 2007; Renoir et al., 2012; Olivier et al., 2013; Shimizu et al., 2013; Sodhi and Sanders-Bush, 2004). Phrasing serotonin and neural plasticity in PubMed showed that one paper was published in 1981, while 72 (2012) and 25 (June 2013) or 30 and 39 (July and September 2013) papers have been published. Hence, the major aim of this book is to examine and summarize recent advances for academia and students. It should be noted, however, that the data commented herein had been mainly observed in adult mammal animals; notwithstanding, some important recent advances in invertebrate species are commented below. Very importantly, as 5-HT receptors may determine occurrence, magnitude, and specificity of plasticity sign on invertebrates and mammals (Kirkwood, 2000), then 5-HT systems could exert multiple functions on memory formation and its alterations (Meneses et al., 2009).

Considering a growing scientific and public interest in mnemonic functions and dysfunctions on humans, it will be of great value that future works attempt to integrate invertebrate and vertebrate studies involving serotonin mnemonic actions. For instance, it is heuristic to look for parallels among species, which might open new avenues to the understanding of neuronal functions and dysfunctions (Meneses et al., 2009; Meneses, 2013).

5-HT systems are involved in memory in different species. For instance, according to Guan et al. (2002) although much is known about short-term integration, little is known about how neurons sum opposing signals for long-term synaptic plasticity and memory storage. In the invertebrate Aplysia, Guam et al. (2012) found that when a sensory neuron simultaneously receives inputs from the facilitatory transmitter 5-HT at one set of synapses and the inhibitory transmitter FMRFamide at another, long-term facilitation is blocked and synapse-specific long-term depression dominates. Guam et al. (2012) reported that chromatin immunoprecipitation assays show that 5-HT induces the downstream gene C/EBP by activating CREB1, which recruits CBP for histone acetylation, whereas FMRFa leads to CREB1 displacement by CREB2 and recruitment of HDAC5 to deacetylate histones. When the two transmitters are applied together, facilitation is blocked because CREB2 and HDAC5 displace CREB1-CBP, thereby deacetylating histones (Guan et al., 2002). Moreover, Rahn et al. (2013) characterize epigenetic mechanisms as critical for the gene expression profile necessary to induce and maintain long-lasting neuronal plasticity and behavior; broadly defined epigenetic mechanisms are a set of processes and modifications influencing gene function without alteration of the primary DNA sequence. Canonical epigenetic mechanisms include histone post-translational modifications (PTMs) and DNA methylation, although recent research has also identified a number of other processes involved in epigenetic regulation, including noncoding RNAs, prions, chromosome position effects, and Polycomb repressors (Rahn et al., 2013). Notably, Jarome and Lubin (2013) highlight that histone lysine methylation is a well-established transcriptional mechanism for the regulation of gene expression changes in eukaryotic cells and is now believed to function in neurons of the central nervous system (CNS) to mediate the process of memory formation and behavior. In mature neurons, methylation of histone proteins can serve to both activate and repress gene transcription. This is in stark contrast to other epigenetic modifications, including histone acetylation and DNA methylation, which have largely been associated with one transcriptional state in the brain. Jarome and

Lubin (2013) discuss the evidence for histone methylation mechanisms in the coordination of complex cognitive processes such as long-term memory (LTM) formation and storage; in addition, the current literature highlights the role of histone methylation in intellectual disability, addiction, schizophrenia, autism, depression, and neurodegeneration (Jarome and Lubin, 2013). Likewise, these authors discuss histone methylation within the context of other epigenetic modifications and the potential advantages of exploring this newly identified mechanism of cognition, emphasizing the possibility that this molecular process may provide an alternative locus for intervention in long-term psychopathologies that cannot be clearly linked to genes or environment alone. Importantly, epigenetic mechanisms, serotonin, and memory seem to have an important link in vertebrate species (Kuhn et al., 2013; Rahn et al., 2013).

Importantly, Monje et al. (2013) reported that Flotillin-1 is an evolutionary-conserved memory-related protein upregulated in implicit and explicit learning paradigms; thus, translational approach from invertebrates to rodents led to the identification of Flotillin-1 as evolutionary-conserved memory-related protein (Monje et al., 2013). Hawkins (2013) hypothesized that in Aplysia, spontaneous release is enhanced by the activation of presynaptic serotonin receptors, but presynaptic D1 dopamine receptors or nicotinic acetylcholine receptors could play a similar role in mammals; similar plasticity occurs in mammals, where it may contribute to reward, memory, and their dysfunctions in several psychiatric disorders.

Certainly, in the last few years a growing number of papers had appeared to directly or indirectly implicate 5-HT systems in learning and memory in species ranging from humans to invertebrates (e.g., see Rajasethupathy et al., 2012; for the original paraphrasing, see Meneses and Perez-Garcia, 2007), and this trend continues (see below). An important insight is provided by PubMed showing that in 1960 two papers were published while 354 publications appeared in 2012, with a peak (370 papers) in 2008, 334 until December 2013, and 3 for publication in 2014. Notwithstanding a limited number of selective 5-HT receptor agonists and antagonists, growing evidence indicates that 5-HT serves as a link between synaptic plasticity at receptor and post-receptor levels (i.e., signal transduction pathways) during learning and memory formation in mammals (Meneses et al., 2009). The above evidence is not only indicating numbers but also quality.

CHAPTER 2

5-HT Systems and Neurobiological Markers Related to Memory Systems

The identification of 5-HT_1 to 5-HT_7 receptor families and its transporter in mammalian species and drugs selective for these sites (Fink and Göthert, 2007; Hoyer et al., 1994) have allowed to dissect their participation in learning and memory (Dougherty and Oristaglio, 2013; King et al., 2008; Puig and Gulledge, 2011; Rodríguez et al., 2012; Terry et al., 2008). Importantly, some 5-HT drugs may present promnesic and/or antiamnesic effects (Table 2.1). Herein, it should be highlighted perhaps an important advantage of 5-HT is that it has diverse pharmacological and genetic tools, neurotoxins, receptor agonist and antagonists (see below), and a well-studied signaling and synaptic modulation in mammal species (Bockaert et al., 2006; Lesch and Waider, 2012). Growing evidence indicates that 5-HT receptors and serotonin transporter (SERT) are involved in normal, pathophysiological, and therapeutic aspects of learning and memory (Gacsályi et al., 2013; Meneses, 1999, 2013). However, 5-HT has been linked to emotional and motivational aspects of human behavior, including anxiety, depression, and impulsivity (Dayan and Huys, 2009; Francis et al., 2010); hence, whether the role of serotonin is related to memory and/or behavioral/emotional aspects (Borroni et al., 2010) remains an important question (for review, see Meneses and Liy-Salmeron, 2012). Evidence revised herein is supporting the former.

For instance, Batsikadze et al. (2013) have observed that beyond its clinical antidepressant effects, serotonin improves motor performance, learning and memory formation; these effects might at least be partially caused by the impact of serotonin on neuroplasticity, which is thought to be an important foundation of the respective functions. These authors highlight an involvement of different mechanisms, including desensitization and downregulation of receptors, or reduction of serotonin synthesis in the effects of chronic administration of selective serotonin reuptake inhibitors (SSRIs) (Batsikadze et al., 2013).

As already mentioned, whether 5-HT markers (e.g., receptors) directly or indirectly participate and/or contribute to the physiological

Table 2.1 5-HT Systems and Neurobiological Markers Related to Memory Systems

Functions Dysfunctions and effects	5-HT Marker	References
Memory formation, aging, AD, and amnesia	SERT, 5-HT$_{1A-1D}$, 5-HT$_{2A/2C}$, 5-HT$_4$, 5-HT$_6$, and 5-HT$_7$ receptors↓	Chou et al. (2012), Eppinger et al. (2011), Meneses (1999, 2003), Meneses and Perez-Garcia (2007), Rodríguez et al. (2012), Xu et al. (2012)
Memory deficits, promnesic and antiamnesic drugs	Modify 5-HT receptors and SERT	Belcher et al. (2005), Eriksson et al. (2012a–c), Garcia-Alloza et al. (2004), Huerta-Rivas et al. (2010), Jones and Moller (2011), Lorke et al., (2006), Marcos et al. (2006, 2008), Mathur and Lovinger (2012), Meneses (2007a,b), Meneses et al. (2011a), Perez-Garcia et al. (2006), Perez-Garcia and Meneses (2008a, 2009), Tellez et al. (2010)
Serotonergic manipulations alter memory	5-HT$_{1A}$, 5-HT$_{2A/2C}$, 5-HT$_3$, 5-HT$_4$, 5-HT$_5$, 5-HT$_6$, and 5-HT$_7$ receptors	Hindi Attar et al. (2012), Bockaert et al. (2008), Gonzalez et al. (2013), Hauder et al. (2012), Meneses (1999, 2003), King et al. (2008), Ögren et al. (2008), Roth et al. (2004), Terry et al. (2008)
Agonists and/or antagonists seem to have promnesic and/ or antiamnesic effects	5-HT$_{1A}$, 5-HT$_4$, 5-HT$_6$, and 5-HT$_7$ receptors	Bockaert et al. (2008), Elvander-Tottie et al. (2009), Ivachtchenko and Ivanenkov (2012), Meneses (1999, 2003), King et al. (2008), Ögren et al. (2008), Roth et al. (2004), Ruiz and Oranias (2010), Terry et al. (2008), van Praag (2004), Youn et al. (2009)

and pharmacological basis of memory and its pathogenesis is a timely question. They seem to be important neurobiological markers (Meneses, 2013). Notably, disorders such as Alzheimer's disease (AD) and schizophrenia have an important component of dysfunctional memory and their etiology includes dysfunction of cholinergic, gluta-matergic, and serotonergic systems (Meneses, 2014; Rodríguez et al., 2012; Terry et al., 2008), and certainly 5-HT has been also implicated in diseases with memory disorders, including depression, compulsive disorder, and posttraumatic stress disorder (PTSD) (Meneses, 2014; Millan et al., 2012; van Praag, 2004; Wallace et al., 2011). Notably, memory disorders appear in diverse diseases, e.g., PTSD, drug addiction and relapse, post-stroke cognitive dysfunctions, schizophrenia, Parkinson disease, and infection-induced memory impairments (Richetto et al., 2013). Of course, future works should clarify these memory dysfunctions, including delimitations.

CHAPTER 3

5-HT Pathways, Receptors and Transporter: Memory Functions and Dysfunctions

5-HT pathways project to almost all brain areas (Jacobs and Azmitia, 1992; Hoyer et al., 1994; Steinbush, 1984) and diverse 5-HT mechanisms might be useful in the treatment of learning and memory dysfunctions. Aging, AD, and amnesia are associated to decrements in 5-HT markers such as SERT and in the number of $5\text{-HT}_{1A/1B}$, $5\text{-HT}_{2A/2C}$, 5-HT_4, 5-HT_6, and 5-HT_7 receptors (Rodríguez et al., 2012; Xu et al., 2012). Importantly, emerging evidence also indicates that memory formation, amnesia, promnesic and amnesic drugs modify serotonergic markers, including 5-HT receptors, SERT, and serotonergic tone (Belcher et al., 2005; Callaghan et al., 2012; Da Silva Costa-Aze et al., 2012; Eriksson et al., 2008, 2012a; Fournet et al., 2012; Garcia-Alloza et al., 2004; Haahr et al., 2012; Hermann et al., 2012; Huerta-Rivas et al., 2010; Jones and Moller, 2011; Lorke et al., 2006; Malá et al., 2013; Marin et al., 2012b; Marcos et al., 2006, 2008; Marshall and O'Dell, 2012; Mathur and Lovinger, 2012; Meneses, 2007a, 2007b; Meneses et al., 2011a; Na et al., 2012; Perez-Garcia et al., 2006; Perez-Garcia and Meneses, 2008a, 2009; Tellez et al., 2010, 2012a, 2012b). Importantly, amnesia and forgetting differ in mechanistic pharmacological and neuroanatomical terms (Tellez et al., 2012b).

Diverse techniques have been used in the identification of serotonergic and/or other neurotransmitter (Ersche et al., 2011; Meyer, 2012) alterations as markers of cognitive processes, including memory formation and memory disorders, ranging from autoradiography, real-time polymerase chain reaction (RT-PCR), Western blot, etc. (for review, see Meneses, 2013; Meneses and Liy-Salmeron, 2012). Even diverse serotonergic manipulations alter memory (Hindi Attar et al., 2012), certainly, 5-HT_{1A}, $5\text{-HT}_{2A/2C}$, 5-HT_3, 5-HT_4, 5-HT_6, and 5-HT_7 receptors have attracted more scientific interest regarding memory (Bockaert et al., 2008, 2011; Cowen and Sherwood, 2013; Meneses, 1999, 2003, 2013; King et al., 2008; Ögren et al., 2008; Roth et al., 2004; Terry et al., 2008). Particularly, $5\text{-HT}_{1A/1B}$, $5\text{-HT}_{2A/2C}$, 5-HT_3, 5-HT_4,

5-HT$_6$, and 5-HT$_7$ receptors have in common that their agonists and/or antagonists seem to have promnesic and/or antiamnesic effects (Bockaert et al., 2008; Borg, 2008; Bombardi and Di Giovanni, 2013; Elvander-Tottie et al., 2009; Hajjo et al., 2012; Ivachtchenko and Ivanenkov, 2012; Ivachtchenko et al., 2012b; King et al., 2008; Meneses, 1999, 2003; Ögren et al., 2008; Pennanen et al., 2013; Roth et al., 2004; Ruiz and Oranias, 2010; Sawyer et al., 2012; Terry et al., 2008; van Praag, 2004; Youn et al., 2009; Yun and Rhim, 2011a,b; Table 2.1). The above apparent paradox remains to be explored, but before continuing discussing 5-HT systems and memory, it is important to mention the problem of protocols of training/testing, memory tasks used, and convergent and divergent findings among laboratories (Meneses, 2013).

3.1 PROTOCOLS OF TRAINING/TESTING, MEMORY TASKS, AND DRUGS

Considering that the number of behavioral memory tasks is abundant (Lynch, 2004; Myhrer, 2003; Peele and Vincent, 1989; Figure 3.1).

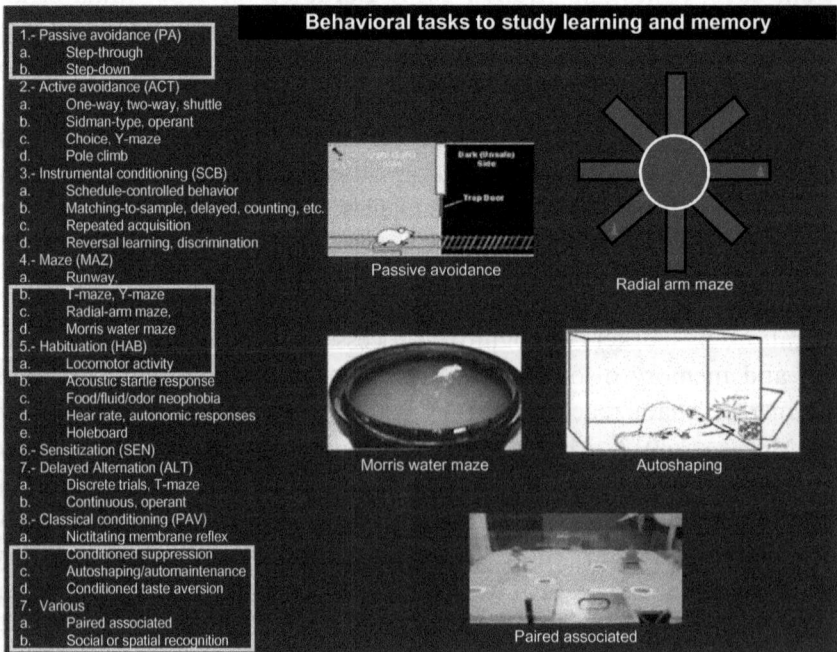

Behavioral tasks to study learning and memory

1.- Passive avoidance (PA)
a.   Step-through
b.   Step-down
2.- Active avoidance (ACT)
a.   One-way, two-way, shuttle
b.   Sidman-type, operant
c.   Choice, Y-maze
d.   Pole climb
3.- Instrumental conditioning (SCB)
a.   Schedule-controlled behavior
b.   Matching-to-sample, delayed, counting, etc.
c.   Repeated acquisition
d.   Reversal learning, discrimination
4.- Maze (MAZ)
a.   Runway,
b.   T-maze, Y-maze
c.   Radial-arm maze,
d.   Morris water maze
5.- Habituation (HAB)
a.   Locomotor activity
b.   Acoustic startle response
c.   Food/fluid/odor neophobia
d.   Hear rate, autonomic responses
e.   Holeboard
6.- Sensitization (SEN)
7.- Delayed Alternation (ALT)
a.   Discrete trials, T-maze
b.   Continuous, operant
8.- Classical conditioning (PAV)
a.   Nictitating membrane reflex
b.   Conditioned suppression
c.   Autoshaping/automaintenance
d.   Conditioned taste aversion
7. Various
a.   Paired associated
b.   Social or spatial recognition

Passive avoidance

Radial arm maze

Morris water maze

Autoshaping

Paired associated

Figure 3.1 Behavioral tasks and memory (modified of Peele and Vincent, 1989).

Rationale for using behavioral tasks rests on the notion that the hippocampus (and rhinal cortices) are paramount for memory formation; in consequence, the cognitive and behavioral requirements are illustrated as well as the differential complexity (based on the brain areas implicated) (Figure 3.2). Herein, we are briefly mentioning three memory tasks used for the investigation of 5-HT systems. In this context, an important issue is convergent and divergent findings among laboratories using similar or different memory tasks, animals employed, etc. (Meneses, 2013). Certainly, it is a complex and multifocal issue; hence, we are briefly addressing a few aspects. Diverse animal models allow the investigation of the various aspects of the memory and its dysfunctions, providing convergent validation of the research findings. Likewise, analysis of protocols of training/testing, memory tasks, and drugs remains as a crucial issue as well as using better devices for measuring memory (Gonzalez et al., 2013; Mar et al., 2013). Indeed, improving behavioral memory tasks and instruments for measuring memory is also important. For instance, new instruments for measuring behavior in autoshaping memory tasks are addressing issues such as variability inter-laboratories, inter-memory tasks, inter-subjects (Cook et al., 2004; Gonzalez et al., 2013; Meneses, 2003) representing

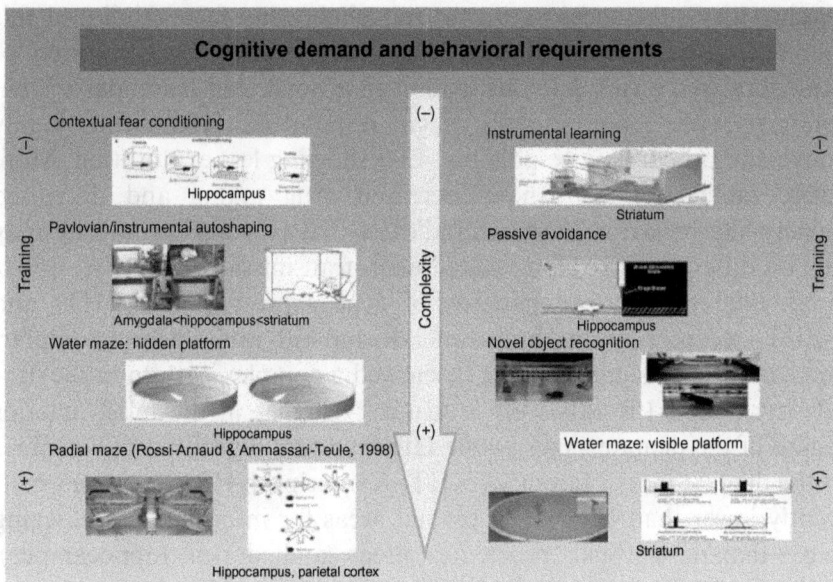

Figure 3.2 Rationale for using behavioral tasks rests on the notion that the hippocampus (and rhinal cortices) are paramount for memory formation. Hence, the cognitive and behavioral requirements are illustrated. The differential complexity is indicated (based on the brain areas implicated).

significant advances (Bussey et al., 2012; Horner et al., 2013; Mar et al., 2013; Markou et al., 2013). Notably, Vanover et al. (2004) reported that the clozapine (which displays affinity for different receptors, see e.g., Meneses, 2014) failed to reach statistical significance due to individual variability but not a $5-HT_{2A}$ receptor inverse agonist neither haloperidol (see Vanover et al, 2004). Of course, the interaction of drug and behavioral task results a heuristic issue for investigating. Doubtless, the intersubject variability is a biological feature, which might be reduced by improving the instruments for measuring memory and/or as Gallistel (2009) suggests that a general solution is a sensitivity analysis: compute the odds for or against the null as a function of the limit(s) on the vagueness of the alternative. If the odds on the null approach 1 from above as the hypothesized maximum size of the possible effect approaches 0, then the data favor the null over any vaguer alternative to it. This same author suggests that the simple computations and the intuitive graphic representation of the analysis (Gallistel, 2009). Otherwise, in the following paragraphs important studies about individual differences are mentioned (Horner et al., 2013; Meneses, 2013).

For instance, Ballaz et al. (2007) reported that sensation seeking is a human personality trait associated with a greater propensity to use psychoactive substances and a rat model showing face validity of this human trait has been developed. The model is based on the variety of behavioral responses that rats exhibit in a novel and inescapable environment with some animals (high responders, HR) being highly active and others (low responders, LR) showing less exploration. More active rats (HR) also show increased drug taking and decreased anxiety-like behavior; evidence indicates that response to novelty may rely on differential 5-HT-mediated neurotransmission. Likewise, Ballaz et al. (2007) compared patterns of gene expression for $5-HT_6$ and $5-ZHT_7$ receptors in the brains of HR and LR rats. Phenotype differences in mRNA signal for $5-HT_6$ showed a complex pattern in the dentate gyrus and LR rats were statistically higher in the most anterior region of the dentate gyrus, while HR rats were higher in median areas of the dentate gyrus. Levels of $5-HT_7$ transcript in HR rats were significantly lower than LR rats in pivotal areas for information trafficking (e.g., thalamo-cortical projection areas and dorsal hippocampus). Ballaz et al. (2007) conclude that their results provide new insight into the potential contribution of 5-HT to novelty seeking behavior and associated behaviors such as substance abuse.

In addition, Flagel et al. (2008) reported that when a discrete cue (a "sign") is presented repeatedly in anticipation of a food reward, the cue can become imbued with incentive salience, leading some animals to approach and engage it, a phenomenon known as "sign tracking" (the animals are sign trackers, STs); in contrast, other animals do not approach the cue, but upon cue presentation they go to the location where food will be delivered (the goal). These animals are known as goal trackers (GTs). It has been hypothesized that individuals who attribute excessive incentive salience to reward-related cues may be especially vulnerable to develop compulsive behavioral disorders, including addiction. Flagel et al. (2008) investigated whether individual differences in the propensity to sign track are associated with differences in responsivity to cocaine; using an autoshaping procedure in which lever (conditioned stimulus, CS) presentation was immediately followed by the response-independent delivery of a food pellet (unconditioned stimulus, US), rats were first characterized as STs or GTs and subsequently studied for the acute psychomotor response to cocaine and the propensity for cocaine-induced psychomotor sensitization. Flagel et al. (2008) found that GTs were more sensitive than STs to the acute locomotor activating effects of cocaine, but STs showed a greater propensity for psychomotor sensitization upon repeated treatment. These data suggest that individual differences in the tendency to attribute incentive salience to a discrete reward-related cue, and to approach and engage it, are associated with susceptibility to a form of cocaine-induced plasticity that may contribute to the development of addiction (Flagel et al., 2008). According to Olshavsky et al. (2013), when pairing a cue with a reward, animals can exhibit both cue-directed ("sign tracking") and reward-directed ("goal tracking") conditioned responses (CRs) and individual differences in the preference for either type of response have long been reported; however, the influence of these phenotypes on responding in aversive tasks is less clear, and their data show that both individuals' propensity for cue-directed orienting and the intensity of the US interact to impact the maintenance of the conditioned fear response (Olshavsky et al., 2013).

3.1.1 Morris Water Maze

Probably, one of the more frequently used behavioral tasks in memory investigation is the water maze. In the water maze (aversively motivated and multiple trials task; Morris, 1984), the experimenter investigates how rodents swim to an escape platform that can be hidden (spatial version of the test) or visible (nonspatial version) (Gerlai, 2001; Myhrer, 2003). The water maze has become a valuable tool for the analysis of mutation effects on brain function, particularly hippocampal function. The water maze has been shown to be able to dissociate hippocampal function (measured in the spatial task) from nonhippocampal function, e.g., general behavioral performance abilities (tested in the nonspatial task) (Gerlai, 2001; Izquierdo et al., 2006). It should be noted here that concerning the water maze, a mechanistic approach to individual differences in spatial learning, memory, and navigation had been reported (Shelton et al., 2014).

3.1.2 Passive Avoidance

One-trial step-down inhibitory (passive) avoidance task in rats (Izquierdo et al., 1999) is an aversively motivated task with very large tradition in memory; and the reasons for using it (Izquierdo et al., 1999) include. First, passive avoidance has a rapid acquisition (seconds), which facilitates the analysis of the time of occurrence of posttraining events. Second, it depends on the integrated activity of CA1, the entorhinal cortex (EC) and the posterior parietal cortex modulated early on by the amygdala and the medial septum. Third, it is the task whose pharmacology and molecular basis have been most extensively studied, particularly in CA1 and EC. Fourth, unlike multi-trial tasks, it permits discrimination between the pharmacology of immediate memory (or working memory, WM) and that of short-term memory (STM). Fifth, using a 0.3–0.4 mA training shock one can obtain retention test latencies far enough from a floor or a ceiling, and therefore easily amenable to the comparative analysis of stimulant and depressant posttraining treatments (Izquierdo et al., 1999). Finally, it has been reliably shown to depend on the actual inhibition of one particular response (stepping down with the four paws on the grid) and not of others (rearing, exploration, sticking the head out, placing just the forepaws on the grid) (Izquierdo et al., 1999).

3.1.3 Autoshaping

Vanover and Barrett (1998) used an autoshaping (or sign tracking) procedure in mice and nose-poke responses into a recessed area, and differentiated by response-dependent reinforcement during two identical consecutive daily sessions. Performance during the first session was considered to be a measure of acquisition and that during the second session a measure of retention. Sensitivity to procedural manipulation, as well as an index of learning under these conditions, was demonstrated, e.g., by a decrease in response rate when nose-poke responses did not produce a reinforcer (Vanover and Barrett, 1998). In addition, the sensitivity of the autoshaping to pharmacological intervention was examined after drug administration before the first session (Vanover and Barrett, 1998). The cholinergic antagonist scopolamine (0.1–10.0 mg/kg) had no effect on acquisition but caused a significant dose-related impairment of retention and the glutamatergic antagonist dizocilpine (0.01–1.0 mg/kg) impaired both acquisition and retention performance. The dual 5-HT$_{1A/7}$ receptor agonist 8-hydroxy-2-(di-n-propylamino)tetralin (8-OH-DPAT; 0.1–1.0 mg/kg) disrupted behavior in general, but failed to have a selective effect on acquisition or retention. Linopirdine (0.1–1.0 mg/kg) showed only a weak enhancement of acquisition, whereas 4-aminopyridine (4-AP; 0.1–1.0 mg/kg) significantly facilitated acquisition (Vanover and Barrett, 1998). Vanover and Barrett (1998) concluded that this paradigm offers the potential for a rapid, objective, and reliable indication of whether a drug will affect the acquisition or retention of a positively reinforced response in mice and could be a useful supplement to existing procedures. In addition, Vanover et al. (2004) reported the pharmacological characterization of AC-90179, a selective 5-HT$_{2A}$ receptor inverse agonist (see below).

Pavlovian/instrumental autoshaping learning task (Vanover and Barrett, 1998) might be conceptualized as an instance of system processing styles: stimulus–stimulus, stimulus–response, and stimulus–reinforcer learning (Meneses, 2003; Meneses et al., 2009), representing a memory task of self-taught situation, which requires brain areas such as dentate gyrus, hippocampal CA1 (Thomas and Everitt, 2001), basolateral amygdaloid nucleus, and PFC (Gonzalez et al., 2013; Perez-Garcia and Meneses, 2008a,b; Tellez et al., 2010, 2012a,b); notably, Pavlovian autoshaping involving frontal cortex (Tomie et al., 2003) and neurobiological findings in Pavlovian/instrumental autoshaping are discussed in the following lines.

3.2 EVIDENCE RELATING 5-HT RECEPTORS AND SERT TO MEMORY

Of the emergent 5-HT systems regarding memory (Figure 3.3), investigation has revealed that the administration of the serotonin precursor

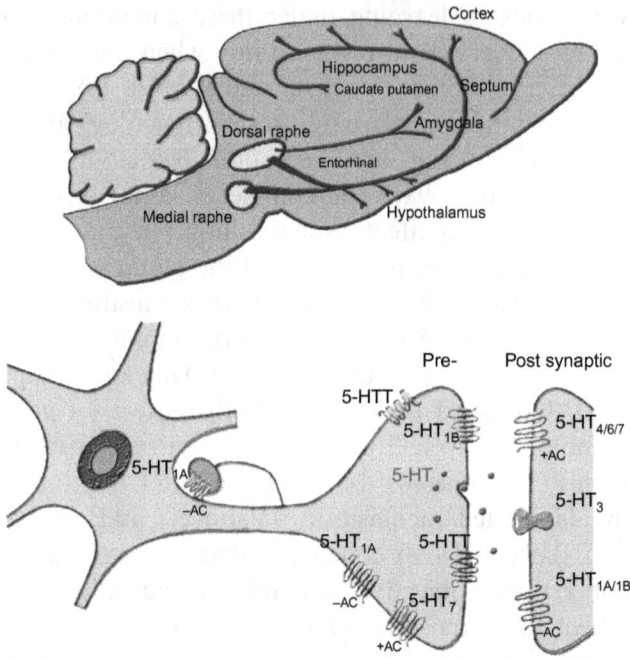

Figure 3.3 Serotonergic projections and synapses.

L-tryptophan (a non dose-dependent manner) facilitated memory consolidation in Pavlovian/instrumental autoshaping (Gonzalez et al., 2013). Importantly, regarding the method of acute tryptophan depletion (ATD), which reduces the availability of the essential amino acid tryptophan (TRP), the dietary serotonin (5-HT) precursor has been applied in many experimental studies. ATD application leads to decreased availability of TRP in the brain and its synthesis into 5-HT. It is therefore assumed that a decrease in 5-HT release and subsequent blunted neurotransmission is the underlying mechanism for the behavioral effects of ATD. However, direct evidence that ATD decreases extracellular 5-HT concentrations is lacking (van Donkelaar et al., 2011). According to van Donkelaar et al. (2011), several studies provide support for alternative underlying mechanisms of ATD; this may question the utility of the method as a selective serotonergic challenge tool. As ATD is used for investigating the role of 5-HT in cognitive functions and psychiatric disorders, the potential of alternative mechanisms and possible confounding factors should be taken into account (van Donkelaar et al., 2011).

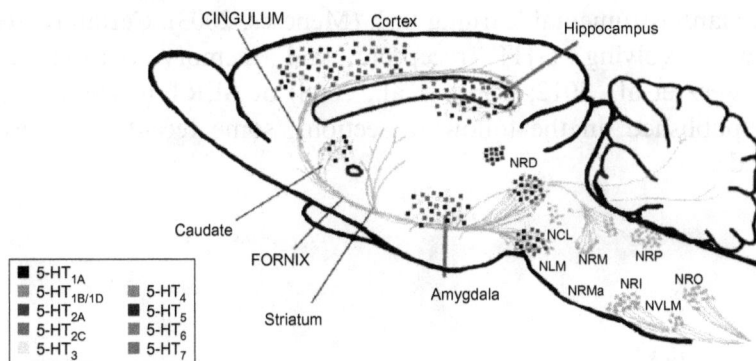

Figure 3.4 5-HT receptors and transporter in rat brain.

Moreover, direct participation of 5-HT has been demonstrated inasmuch as enhanced brain serotonin activity by means of its precursor (i.e., tryptophan) improved memory in animals (Haider et al., 2012) and in normal elderly people and AD patients and schizophrenics (Golightly et al., 2001; Porter et al., 2003), whereas human and animals decreasing brain 5-HT levels by acute 5-HT depletion impaired it (Schmitt et al., 2006). This evidence is consistent with the results that posttraining (but not pre-) administration of 5-HT uptake inhibitors improved memory consolidation by using multiple 5-HT receptors (Meneses, 2002, 2003) distributed pre- and postsynaptically in brain areas related to memory (Figures 3.3 and 3.4).

It should be noted that ATD-induced reduction in somatodendritic 5-HT$_{1A}$ autoreceptor binding (ligand [^{3}H]-WAY100635) may represent an intrinsic "homeostatic response" reducing serotonergic feedback in dorsal raphe projection areas without postsynaptic alterations (frontal cortex, remaining cortex, and hippocampus). In contrast, the increase in 5-HT$_{2A}$ receptor (ligand [^{3}H]ketanserin) after chronic tryptophan depletion (CTD) may be a compensatory response to a long-term reduction in 5-HT (Cahir et al., 2007, 2008); likewise, ATD does not alter central or plasma brain-derived neurotrophic factor in the rat in midbrain and hippocampus (Cahir et al., 2008).

But the effects of 5-HT endogenous on memory formation, using a 5-HT uptake facilitator (tianeptine) or inhibitor (fluoxetine) and selective 5-HT$_{1-7}$ receptor antagonists, have showed that posttraining tianeptine injection enhanced memory consolidation in an autoshaping

Pavlovian/instrumental learning task (Meneses, 2003). Certainly, recent reviews involving 5-HT receptors and memory (Cassel, 2010; Rodríguez et al., 2012; Terry et al., 2008) or SERT in memory have been published; in the following sections, some recent advances are revised.

CHAPTER 4

5-HT$_{1A}$ Receptor

Probably 5-HT$_{1A}$ receptor is one of the most investigated 5-HT receptors (Calcagno et al., 2006; Carli and Samanin, 1992, 2000; Carli et al., 1995, 1997, 1998, 1999a,b, 2001, 2006; Eriksson et al., 2012a; Haider et al., 2012; Lladó-Pelfort et al., 2012; Millan et al., 2004; Meneses and Perez-Garcia, 2007; Ögren et al., 2008; Schechter et al., 2004, 2005), which allows a multifocal analysis (likewise allowing a more rich analysis of 5-HT$_2$, 5-HT$_4$, 5-HT$_6$, and 5-HT$_7$ receptors as well as SERT, see below). Indeed, evidence indicates that 5-HT$_{1A}$ receptor agonists and antagonists have promnesic or antiamnesic effects. Notably, partial agonists of this receptor (e.g., tandospirone; Sumiyoshi et al., 2008) might be useful in the treatment of memory alterations in schizophrenia.

Moreover, Izquierdo et al. (1998) had studied STM and LTM. Indeed, rats with cannulae implanted in the dorsal CA1 region of the hippocampus or in the EC were trained in one-trial step-down inhibitory avoidance and tested 1.5 h (STM) or 24 h (LTM) later (Izquierdo et al., 1998). Drugs infused immediately posttraining inhibited STM without altering LTM: the 5-HT$_{1A/7}$ receptor agonist 8-OH-DPAT given into CA1 and the 5-HT$_{1A}$ antagonist NAN-190 given into EC. According to Izquierdo et al. (1998), these findings indicate that STM is not a necessary step toward LTM. In addition, intra-entorhinal 8-OH-DPAT enhanced STM and depressed LTM; none of the drugs had any effect on retrieval of either STM or LTM when given prior to testing. The data indicate that STM and LTM are differentially modulated by 5-HT$_{1A}$ receptors in CA1 and EC (Izquierdo et al., 1998). Moreover, now it is clear that the 5-HT$_{1A/7}$ receptor agonist 8-OH-DPAT at high doses impaired memory in memory tasks such as water maze and Pavlovian/instrumental autoshaping, while at low doses it improved performance and/or had antiamnesic effects; the blockade of 5-HT$_{1A}$ receptor had no effect but can reverse amnesia induced by cholinergic or glutamatergic agents (e.g., see Meneses, 2007a,b; for review, see Meneses and Perez-Garcia, 2007; see also Hirst et al., 2008; Ögren et al., 2008; Wedzony et al., 2000). In addition, administered into the dorsal raphe 8-OH-DPAT had no effect on water maze but compensated the deficit on spatial learning caused by impaired cholinergic or glutamatergic hippocampal transmission (Carli et al., 2001) and enhanced operant conditional discrimination (Ward et al., 1999). On operant autoshaping task, mice lacking $5-HT_{1A}$ or $5-HT_{1B}$ receptors learned faster (Pattij, 2002), and even the latter mice showed water maze enhanced platform acquisition and probe trial performance. Now it is clear that 8-OH-DPAT is a dual $5-HT_{1A/7}$ receptor agonist, which even displays affinity for other neurotransmission systems.

8-OH-DPAT displays affinity for the $5-HT_{1A}$ and $5-HT_7$ receptors (Gobert et al., 1995; Matthys et al., 2011; Sharp et al., 1989) and functional activity via α_1-adrenoceptors (Castillo et al., 1993). This information is consistent with the evidence that in part, $5-HT_7$ receptor mediates the 8-OH-DPAT memory effects in autoshaping (Manuel-Apolinar and Meneses, 2004; Perez-Garcia and Meneses, 2009) as well as in passive avoidance (Eriksson et al., 2012a; Horisawa et al., 2013; Izquierdo et al., 1999). It should be noted that rats in Pavlovian autoshaping paired had high CR, which showed more rapid CR acquisition and higher asymptotic levels of lever-press autoshaping CR performance relative to rats in low CR group (Tomie et al., 2003). The omission group received autoshaping with an omission contingency, such that performing the lever-press autoshaping CR resulted in the cancellation the food US, while random group received presentations of lever CS and food US randomly with respect to one another. Though omission and random groups showed poor scores of CR (relative to Pavlovian autoshaping) and did not differ in lever-press autoshaping CR performance, omission group showed significantly lower levels of ^{3}H-8-OH-DPAT-labeled $5-HT_{1A}$ binding in postsynaptic areas (frontal cortex, septum, caudate putamen), as well as significantly higher plasma corticosterone levels than random group (Tomie et al., 2003). In addition, random group showed higher levels of ^{3}H-8-OH-DPAT-labeled $5-HT_{1A}$ binding in presynaptic (somatodendritic autoreceptors, dorsal raphe nucleus (RN)) relative to each of the other three groups. Also, autoradiographic analysis of ^{125}I-LSD-labeled $5-HT_{2A}$ receptor binding revealed no significant differences between paired high CR and paired low CR groups or between omission and random groups in any brain regions (Tomie et al., 2003). It should be noted that regardless the level of CR, both paired groups had the higher expression of $5-HT_{1A}$ (presynaptic) and/or $5-HT_{2A}$ (postsynaptic) receptors (Meneses et al., 2004a). Notably, Jen et al. (2008) reported that downregulation of the 5-HT system in the limbic system, i.e., the reduction of the hippocampus 5-HT content and the amygdala 5-HT$_{1A}$ receptor expression, may be involved in the exercise-enhanced fear memory. Interestingly, Tsetsenis et al. (2007) reported that 5-HT$_{1A}$ receptor 1A knockout (Htr1a(KO)) mice show increased fear conditioning specifically to partially conditioned cues arguing that these animals harbor a defect in the cognitive processing of aversive cues.

Harvey et al. (2003) studied the effects of a time-dependent sensitization (TDS) model of stress on spatial memory deficits, 1 week poststress, using the water maze, determining basal and seven-day poststress plasma corticosterone levels. In addition, due to the putative role of serotonin in anxiety and stress, and in the treatment of PTSD, hippocampal 5-HT$_{1A}$ and PFC 5-HT$_{2A}$ radioligand binding studies were performed. Harvey et al. (2003) found that TDS stress evoked a marked deficit in spatial memory on day 7 post-TDS stress, coupled with significantly depressed plasma corticosterone levels, and cognitive and endocrine changes at day 7 poststress were associated with a significant increase in receptor density (B(max)) and a significant decrease in receptor affinity (K(d)) for hippocampal 5-HT$_{1A}$ receptors. The B(max) of PFC 5-HT$_{2A}$ receptors was unaffected, but K(d) was significantly increased. Harvey et al. (2003) conclude that TDS stress evokes cognitive and endocrine changes characteristic of PTSD and TDS stress, induces diverse adaptive 5-HT receptor changes in critical brain areas involved in emotion and memory that may underlie the effect of stress on cognitive function.

Studies involving expression of 5-HT$_{1A}$, 5-HT$_{2A}$, 5-HT$_4$, and 5-HT$_6$ receptors in Pavlovian/instrumental autoshaping memory formation, amnesia conditions (e.g., pharmacological models or aging), and recovery of amnesia (Perez-Garcia and Meneses, 2008a) showed that specific 5-HT receptors were expressed in trained animals relative to untrained in brain areas such as cortex, hippocampus, and amygdala. However, relative to the control group, rats with amnesia or recovered of memory showed in the hippocampus region where explicit memory is formed, a complex pattern of 5-HT receptor expression. An intermediate expression occurred in amygdala, septum, and some cortical areas in charge of explicit memory storage. In brain areas thought to be in charge of procedural memory such as basal ganglia, animals showing recovered memory displayed an intermediate expression, while amnesic groups, depending on the pharmacological amnesia model, showed up- or downregulation. In conclusion, evidence indicates that autoradiography (using specific radioligands) offers excellent opportunities to map dynamic changes in brain areas engaged in these cognitive processes. The 5-HT modulatory role strengthens or suppresses memory is critically depend on the timing of the memory formation (Perez-Garcia and Meneses, 2008a).

How functional are these receptors? The 8-OH-DPAT facilitatory effect on autoshaped memory (LTM 24 and 48 h) was accompanied by cAMP (cyclic adenosine monophosphate) increased cortical and hippocampal cAMP production (Manuel-Apolinar and Meneses, 2004; Meneses et al., 2002). The selective antagonists WAY100635 ($5\text{-}HT_{1A}$) or DR4004 ($5\text{-}HT_7$ receptor; Kikuchi et al., 2003) alone had no effect on memory, but the 8-OH-DPAT−WAY100635 or 8-OH-DPAT−DR4004 combinations did not modify or enhance 8-OH-DPAT-induced increased cAMP, respectively (Manuel-Apolinar and Meneses, 2004).

It should be noted that as Izquierdo et al. (2006) found augmented cAMP production (immediately and 90 min later) and memory formation in passive avoidance task, hence we studied further cAMP and memory (Perez-Garcia and Meneses, 2008a) (see $5\text{-}HT_7$ receptor section).

Finally, the antiphospholipid syndrome (APS) is an autoimmune disease with the presence of high titers of circulating autoantibodies, thrombosis with consecutive infarcts; a mouse model of APS with documented neurological and cognitive deficits was investigated by Frauenknecht et al. (2013). Ligand binding densities of NMDA, AMPA, $GABA_A$, and $5\text{-}HT_{1A}$ receptors were analyzed *in vitro* receptor autoradiography showing that binding values of excitatory and inhibitory neurotransmitter receptors were largely unchanged; however, $5\text{-}HT_{1A}$ receptor in the hippocampus and in the primary somatosensory cortex of eAPS mice were significantly upregulated; and according to Frauenknecht et al. (2013), possibly related to the behavioral abnormalities observed.

Notably, Borg (2008) had highlighted that animal studies and pharmacological studies in man have suggested that the serotonin $5\text{-}HT_{1A}$ receptor may serve as a biomarker for cognitive functioning and a target for treatment of cognitive impairment. Consistent findings in man have nonetheless hitherto remained sparse. Positron emission tomography (PET) imaging of the 5-HT$_{1A}$ receptor in patients with AD, schizophrenia, and depression implicate an alteration in this receptor binding compared to control subjects, but it is yet unknown whether these alterations are related to the cognitive impairment associated with these disorders. Pharmacological challenge studies using 5-HT$_{1A}$ agonism and antagonism to manipulate the serotonin system support involvement of the 5-HT$_{1A}$ receptor in human cognition, mainly in verbal memory functioning (Borg, 2008); however, the effect varies across studies and it remains unclear if the 5-HT$_{1A}$ receptor serves as a validated target for treatment of cognitive deficits. This lack of confirmation of experimental preclinical data calls for increased efforts in translational research. Molecular imaging techniques, such as PET, hold the potential to facilitate translational neuroscience by confirming observations from animal models in man and aid development of validated animal models of use for advancement of pharmacological treatment (Borg, 2008). Furthermore, in combination with molecular genetics, molecular imaging may suggest novel strategies for prevention and intervention based on an understanding of the molecular mechanisms involved in disease pathogenesis of major neuropsychiatric disorder and associated cognitive impairment (see Borg, 2008; see also Newman-Tancredi and Kleven, 2011; also available radioligands for the 5-HT$_{1A}$, 5-HT$_{1B}$, 5-HT$_{2A}$, 5-HT$_4$, and 5-HT$_6$ receptors, see Saulin et al., 2012; Zimmer and Le Bars, 2013). Finally, the inhibitory effects of nicotine and the role of 5-HT$_{1A}$ receptors in this effect seem to be involving different functions via RN (Hernandez-Lopez et al., 2013) including memory. Importantly, there is 5-HT neuron diversity in the dorsal raphe (Andrade and Haj-Dahmane, 2013) and RN, which is composed of serotonergic neurons as well as nonserotonergic neurons, projecting to almost all of the brain. Interestingly, Sumiyoshi et al. (2013) highlight the neural basis for the ability of atypical antipsychotic drugs (e.g., 5-HT$_{1A}$ receptor partial agonists) to improve cognition in schizophrenia.

CHAPTER 5

5-HT$_{1B}$ Receptor

Evidence (Meneses, 2001) indicates that SB-224289 (a 5-HT$_{1B}$ receptor inverse agonist) posttraining injection facilitated memory consolidation in Pavlovian/instrumental autoshaping task. This effect was partially reversed by GR 127935 (a 5-HT$_{1B/1D}$ receptor antagonist), but unaffected by MDL 100907 (a 5-HT$_{2A}$ receptor antagonist) or ketanserin (a 5-HT$_{1D/2A/7}$ receptor antagonist) at low doses. Moreover, SB-224289 antagonized the memory deficit produced by TFMPP (N-(3-trifluoromethylphenyl) piperazine, a 5-HT$_{1A/1B/1D/2A/2C}$ receptor agonist), GR 46611 (a 5-HT$_{1A/1B/1D}$ receptor agonist), mCPP (1-(3-chlorophenyl) piperazine, a 5-HT$_{2A/2C/3/7}$ receptor agonist/antagonist), or GR 127935 (at low dose). SB-224289 did not alter the 8-OH-DPAT (a 5-HT$_{1A/7}$ receptor agonist) memory facilitatory effect but SB-224289 eliminated the deficit memory produced by the anticholinergic muscarinic scopolamine or the glutamatergic antagonist dizocilpine. Administration of both, GR 127935 (5 mg/kg) plus ketanserin (0.01 mg/kg) did not modify memory consolidation; nevertheless, when ketanserin dose was increased (0.1–1.0 mg/kg) and SB-224289 dose was maintained constant, a memory facilitation effect was observed. Notably, SB-224289 at 1.0 mg/kg potentiated a subeffective dose of the 5-HT$_{1B/1D}$ receptor agonist/antagonist mixed GR 127935, which facilitated memory consolidation and this effect was abolished by ketanserin at a higher dose. Collectively, the data confirm and extend the earlier findings with GR 127935 and the effects of nonselective 5-HT$_{1B}$ receptor agonists. Clearly 5-HT$_{1B}$ agonists induced a memory deficit which can be reversed with SB-224289; perhaps more importantly, SB-224289 enhances memory consolidation when given alone and can reverse the deficits induced by both cholinergic and glutamatergic antagonists. Hence, 5-HT$_{1B}$ receptor inverse agonists or antagonists could represent drugs for the treatment of learning and memory dysfunctions (Meneses, 2001).

Buhot et al. (2003b) reported that 5-HT$_{1B}$ receptor KO mice were submitted to various behavioral paradigms carried out in the same experimental context (water maze), aiming at exposing mice to various levels of memory demand. 5-HT$_{1B}$ KO mice exhibited facilitation in the acquisition of a hippocampal-dependent spatial reference memory task in the water maze. This facilitation was selective of task difficulty, showing thus that the genetic inactivation of the 5-HT$_{1B}$ receptor is associated with facilitation when the complexity of the task is increased; a protective effect on age-related hippocampal-dependent memory decline was observed. Young-adult and aged KO and wild-type (WT) mice were equally able to learn a delayed spatial matching-to-sample WM task in a radial-arm water maze with short (0 or 5 min) delays. However, 5-HT$_{1B}$ receptor KO mice, only, exhibited selective memory impairment at intermediate and long (15, 30, and 60 min) delays. Treatment by scopolamine induced the same pattern of performance in wild type as did the mutation for short (5 min, no impairment) and long (60 min, impairment) delays. These data revealed a beneficial effect of the mutation on the acquisition of a spatial reference memory task, but a deleterious effect on a WM task for long delays. This 5-HT$_{1B}$ receptor KO mouse story highlights the problem of the potential existence of "global memory enhancers" (Buhot et al., 2003b).

In addition, Buhot et al. (2003a) compared the performances of young-adult (3 months old) and aged (22 months old) 5-HT$_{1B}$ receptor KO and WT mice in the same task. Young-adult and aged KO mice exhibited facilitated acquisition of the reference memory task as compared to their respective WT controls. Generally, the performance of aged KO was similar to that of young-adult WT on the parameters defining performance and motor (swim speed) aspects of the task. During probe trials, all mice presented a spatial selectivity, which was, however, less pronounced in aged than in young-adult WT. No such age-related effect was observed in KO mice. In a massed spatial learning task, aged KO and WT mice globally exhibited the same level of performance. Nevertheless, young-adult and aged KO mice were superior to their WT controls as concerns the WM component of the task. According to Buhot et al. (2003a), 5-HT$_{1B}$ KO mice are more resistant than WT to age-related memory decline as concerns both reference/long-term and working/short-term spatial memory.

More recently, partial data about 5-HT$_{1B}$ receptor and memory processes (Drago et al., 2010) arising from genetic association studies suggest that the HTR1B gene plays a relevant role in substance-related and obsessive compulsive disorders; and no solid evidence for other psychiatric disorders was found. This finding is quite striking because of the heavy impairment of motivation and of mnemonic-related functions (e.g., recall bias) that characterize major psychiatric disorders (Drago et al., 2010). Moreover, a serotonin neuron-targeting function of glycogen synthase kinase-3β (GSK3β) by regulating 5-HT$_{1B}$ autoreceptors impacts serotonergic neuron firing, serotonin release, and serotonin-regulated behaviors (Zhou et al., 2012). Importantly, altered levels of GSK-3β and β-catenin are associated with various neuropsychiatric and neurodegenerative diseases, while various classical neuropsychiatric drugs inhibit GSK-3β and upregulate β-catenin expression (Wada, 2009). Providing further support to the notion that 5-HT$_{1B}$ receptor plays a role in memory formation, Cai et al. (2013) reported that the 5-HT$_{1B}$ receptor antagonist SB-216641 significantly improved retention of spatial information in the water maze. Woehrle et al. (2013) found that chronic fluoxetine treatment reversed the serotonin 1B receptor-induced deficits in delayed alternation. Eriksson et al. (2008) reported that blockade of 5-HT$_{1B}$ receptor facilitates contextual aversive learning in mice by disinhibition of cholinergic and glutamatergic neurotransmission; notably, Eriksson et al. (2012b) in an elegant experiment found bidirectional regulation of emotional memory involving 5-HT$_{1B}$ receptor and its hippocampal adapter protein p11.

CHAPTER 6

5-HT$_{1E/1F}$ Receptor

The presence of 5-HT$_{1E/FE}$ receptors in brain areas involved in memory suggests that they might have a role in memory. For instance, the stimulation of 5-HT$_{1E}$ receptor and subsequent inhibition of adenylate cyclase activity in the dentate gyrus showed that this receptor may mediate regulation of hippocampal activity by 5-HT, making it a possible drug target for the treatment of neuropsychiatric disorders characterized by memory deficits (e.g., AD) or as a target for the treatment of temporal lobe epilepsy (Klein and Teitler, 2012).

In the case of 5-HT$_{1F}$ receptor, Lucaites et al. (2005) found specific 5-HT$_{1F}$ receptor binding in rat brain of layers 4−5 of all cortical regions examined, as well as olfactory bulb and tubercle, nucleus accumbens (NAc), caudate putamen, parafascicular nucleus of the thalamus, medial mammillary nucleus, the CA3 region of the hippocampus, subiculum, and several amygdaloid nuclei. Some species differences in the distribution of the 5-HT$_{1F}$ receptor were noted and preliminary binding studies in rhesus monkey and human brain sections showed 5-HT$_{1F}$ receptor in cortical layers 4−5, subiculum (in the monkey), and the granule cell layer of the cerebellum (Lucaites et al., 2005) and *in vitro* evidence indicates that 5-HT$_{1F}$ receptor coupled negatively to cAMP production and 5-HT$_{1F}$ receptor is localized in brain areas mediating memory, including cortex, hippocampus, and RN. Preliminary evidence (Ponce-Lopez and Meneses, 2008) determined the effects of 5-HT$_{1F}$ receptor agonist LY344864 (LY) on STM and LTM and cAMP productions. Rats received autoshaping training and immediately afterwards were treated with vehicle (mg/kg), LY (0.01−10), methiothepin (no selective 5-HT antagonist 5-HT), el GR127935 (5-HT$_{1D}$ antagonist) or SB-224289 (5-HT$_{1B}$ antagonist) and tested for STM (1.5 h) and LTM (24−48 h); following STM and LTM rats were euthanized, and PFC, hippocampus, and RN were extracted for the cAMP ELISA immunoassay (Ponce-Lopez and Meneses, 2008). LY did not affect STM; nonetheless, 0.01 and 0.3−10.0 mg/kg impaired LTM (24 h); the amnesic-like effects induced by LY344864 (0.3 mg/kg) were partially reversed by methiothepin, GR127935, or SB-224289 (1), which alone had no effects. LTM memory formation (vehicle group) was associated with increased cAMP production in RN, hippocampus, and PFC (unpublished data; Ponce-Lopez and Meneses, 2009); the amnesia induced by LY344864 was accompanied by increment in cAMP production in RN but decrement in hippocampus and PFC. Hence, changes in cAMP production (via 5-HT$_{1F}$ activation) in the route of signaling of memory formation and amnesia are critical, possibly therapeutic application of selective 5-HT$_{1F}$ receptor antagonist for the treatment of memory dysfunctions related to age or amnesia (Ponce-Lopez and Meneses, 2008). Of course, experiments such as determination if in the effects of LY344864 5-HT$_{1F}$ receptor was the only responsible as well as the 5-HT$_{1F}$ receptor interaction with other serotonergic receptors and/or neurotransmission systems.

CHAPTER 7

5-HT$_{2A/2B/2C}$ Receptor

The analysis of 5-HT$_2$ receptors role in memory consolidation (Meneses, 2002) revealed that the SB-200646 (a selective 5-HT$_{2B/2C}$ receptor antagonist) and LY215840 (a nonselective 5-HT$_{2/7}$ receptor antagonist) posttraining administration had no effect on an autoshaped memory consolidation; however, both drugs significantly and differentially antagonized the memory impairments induced by serotonergic drugs as mCPP, 1-naphthyl-piperazine (1-NP), mesulergine, or TFMPP. In contrast, SB-200646 failed to modify the facilitatory procognitive effect produced by $(+/-)$-2,5-dimethoxy-4-iodoamphetamine (DOI) or ketanserin, which were sensitive to MDL100907 (selective 5-HT$_{2A}$ receptor antagonist) and to a LY215840 high dose. Finally, SB-200646 reversed the memory deficit induced by dizocilpine, but not that by scopolamine: while SB-200646 and MDL100907 coadministration reversed memory deficits induced by both drugs. Hence, 5-HT$_{2B/2C}$ receptors might be involved in memory formation probably mediating a suppressive or constraining action; certainly, whether the drug-induced memory impairments in the above study explained by simple agonism, antagonism, or inverse agonism at 5-HT$_2$ receptors remain unclear (Meneses, 2002). 5-HT$_2$ receptor subtypes blockade may provide some benefit to reverse poor memory consolidation associated with decreased cholinergic, glutamatergic, and/or serotonergic neurotransmission (Meneses, 2002).

As already mentioned, Vanover et al. (2004) reported that AC-90179 (a selective 5-HT$_{2A}$ receptor inverse agonist) had no effect on acquisition of a (Pavlovian/instrumental autoshaping) nose-poke response until the highest dose (30 mg/kg s.c.). Haloperidol significantly reduced the number of reinforcers earned and clozapine dose-dependently decreased the number of reinforcers earned, but the effect failed to reach statistical significance due to individual variability (Vanover et al., 2004). Regarding individual variability in autoshaping tasks; see e.g., Meneses, 2013; for other groups using autoshaping tasks see Perez-Garcia and Meneses (2008a) and some of these data are comparable to those observed in other memory tasks (Perez-Garcia and Meneses, 2008a).

Very importantly, the status of inverse agonism at serotonin 2A (5-HT_{2A}) and 5-HT_{2C} receptors was recently revised (Aloyo et al., 2009). Indeed, contemporary receptor theory was developed to account for the existence of constitutive activity, as defined by the presence of receptor signaling in the absence of any ligand; thus, ligands acting at a constitutively active receptor can act as agonists, antagonists, and inverse agonists (Aloyo et al., 2009). *In vitro* and *ex vivo* studies have also revealed the complexity of ligand/receptor interactions, including agonist-directed stimulus trafficking, a finding that has led to multi-active state models of receptor function (Aloyo et al., 2009; Meneses, 2013). Studies with a variety of cell types have established that serotonin 5-HT_{2A} and 5-HT_{2C} receptors also demonstrate constitutive activity and inverse agonism (Aloyo et al., 2009); however, until recently, there has been no evidence to suggest that these receptors also demonstrate constitutive activity and hence reveal inverse agonist properties of ligands *in vivo*. Aloyo et al. (2009) describe the current knowledge of constitutive activity *in vitro* and then examine the evidence for constitutive activity *in vivo*. According to Aloyo et al. (2009), both 5-HT_{2A} and 5-HT_{2C} receptors are involved in a number of physiological and behavioral functions and are the targets for treatment of schizophrenia, anxiety, weight control, Parkinsonism, and other disorders; the existence of constitutive activity at these receptors *in vivo*, along with the possibility of inverse agonism, provides new avenues for drug development (Aloyo et al., 2009). In the context of memory, 5-HT_{2A} receptor inverse agonists seem to be important.

It should be noted that the inverse agonism and its therapeutic significance postulate that a large number of G-protein-coupled receptors (GPCRs) show varying degrees of basal or constitutive activity. This constitutive activity is usually minimal in natural receptors but is markedly observed in WT and mutated (naturally or induced) receptors (Khilnani and Khilanani, 2011). Conventional two-state drug–receptor interaction model, binding of a ligand may initiate activity (agonist with varying degrees of positive intrinsic activity) or prevent the effect of an agonist (antagonist with zero intrinsic activity) (Khilnani and Khilnani, 2011). Inverse agonists bind with the constitutively active receptors, stabilize them, and thus reduce the activity (negative intrinsic activity). According to Khilnani and Khilanani (2011), receptors of many classes (α-and β-adrenergic, histaminergic, GABAergic, serotonergic, opiate, and angiotensin receptors) have shown basal activity in suitable *in vitro* models. For instance, drugs that have been conventionally classified as antagonists (e.g., β-blockers, antihistaminics) have shown inverse agonist effects on corresponding constitutively active receptors and some drugs have significant inverse agonistic activity that contributes partly or wholly to their therapeutic value (Khilnani and Khilanani, 2011). Inverse agonism may also help explain the underlying mechanism of beneficial effects of clozapine in psychosis (see below). Notably, understanding inverse agonisms has paved a way for newer drug development (Khilnani and Khilnani, 2011), which has only desired therapeutic value and is devoid of unwanted (or reduced) adverse effect (e.g., anxiety, antinociceptive, obesity, chronic asthma). According to Khilnani and Khilanani (2011), pimavanserin (ACP-103), a highly selective 5-HT$_{2A}$ inverse agonist, attenuates psychosis in patients with Parkinson's disease with psychosis and is devoid of extrapyramidal side effects; therefore, inverse agonism is an important aspect of drug–receptor interaction and has immense untapped therapeutic potential (Khilnani and Khilanani, 2011).

Navailles et al. (2013) updating the growing number of studies show (by means of pharmacological tools) the participation of the constitutive activity of 5-HT$_{2C}$ receptors in the control of various biochemical and behavioral functions *in vivo* and emphasize the functional organization of this constitutive control together with the phasic and tonic (involving the spontaneous release of 5-HT) modalities of the 5-HT$_{2C}$ receptor in the brain (Navailles et al., 2013). Moreover, functional anatomy of 5-HT$_{2A}$ receptors in the amygdala and hippocampal complex revealed relevance to memory functions (Bombardi and Di Giovanni, 2013). Investigation about 5-HT$_{2A}$ receptor role in memory had shown (Williams et al., 2002) that memory fields of putative pyramidal cells were attenuated by iontophoresis of 5-HT$_{2A}$ antagonists, which primarily produced a reduction in delay activity for preferred target locations. Conversely, 5-HT$_{2A}$ stimulation by α-methyl-5-HT or 5-HT itself accentuated the spatial tuning of these neurons by producing a modest increase in activity for preferred target locations and/or a reduction in activity for nonpreferred locations. The agonist effects could be reversed by the selective antagonist MDL100907 and were dose dependent, such that high levels attenuated spatial tuning by profoundly reducing delay activity. A role for feed-forward inhibitory circuitry in these effects was supported by the finding that 5-HT$_{2A}$ blockade also attenuated the memory fields of putative interneurons. Williams et al. (2002) conclude that prefrontal 5-HT$_{2A}$ receptors have a hitherto unrecognized role in the cognitive function of WM, which involves actions at both excitatory and inhibitory elements within local circuitry (Zhang et al., 2013). Notably, Dougherty and Oristaglio (2013) hypothesize that long-term drug treatments resulting in 5-HT$_{2A}$ receptor upregulation may be useful in enhancing recall of learned behaviors and thus may have potential for treating cognitive impairment associated with neurodegenerative disorders. Likewise, Dougherty and Oristaglio (2013) point out that their observations suggest a widespread modulatory role of 5-HT$_{2A}$ and 5-HT$_{2C}$ receptors in learning and memory with the net effect being dependent on task requirements and the specific mnemonic systems recruited. It should be noted that the methodology employed by Dougherty and Oristaglio (2013) is a first step in better characterizing drug effects on goal-directed behavior and identifying quantifiable factors that can underlie changes in response latency; they suggest using more elaborate methodologies, such as video tracking, could add resolution to this analysis and provide a more complete profile of motor variability under baseline and drug-influenced conditions. Such analyses could be an important consideration for evaluating the behavioral performance of different strains of mice, particularly in aged or neurodegenerative models where latencies in choice behavior and motor variability might be considerably higher (Dougherty and Oristaglio, 2013). For instance, increases in both reaction time and reaction time variability on cognitive tasks are associated with aging and cognitive decline in humans (Dougherty and Oristaglio, 2013).

Moreover, 5-HT$_{2B/2C}$ receptors and memory investigation (Boulougouris and Robbins, 2010) had shown that systemic administration of 5-HT$_{2C}$ and 5-HT$_{2A}$ receptor antagonists significantly enhanced and impaired spatial reversal learning, respectively (Lopez-Velazquez et al., 2011). Indeed, the role of these receptors in the mediation of these opposing effects was further investigated regarding neuroanatomical specificity within some of the brain regions implicated in cognitive flexibility (Boulougouris and Robbins, 2010), namely the orbitofrontal cortex (OFC), medial prefrontal cortex (mPFC), and NAc by means of targeted infusions of selective 5-HT$_{2C}$ and 5-HT$_{2A}$ receptor antagonists (SB-242084 and M100907, respectively). Intra-OFC 5-HT$_{2C}$ receptor antagonism produced dose-dependent effects similar to those of systemic administration, i.e., improved spatial reversal learning by reducing the number of trials (doses: 0.1, 0.3, and 1.0 µg) and perseverative errors to criterion (0.3 and 1.0 μg) compared with controls (Boulougouris and Robbins, 2010). However, the highest dose (1.0 μg) showed a nonselective effect, as it also affected retention preceding the reversal phase and decreased learning errors. Intracerebral infusions of SB-242084 into the mPFC or NAc produced no significant effects on any behavioral measures. Similarly, no significant differences were observed with intra-OFC, -mPFC, or -NAc infusions of M100907. According to Boulougouris and Robbins (2010), these data suggest that the improved performance in reversal learning observed after 5-HT$_{2C}$ receptor antagonism is mediated within the OFC. Also, the data also bear on the issue of whether 5-HT$_{2C}$ receptor antagonism within the OFC might help elucidate the biological substrate of obsessive−compulsive disorder offering the potential for therapeutic application. Moreover, novel 5-HT$_{2A/2C}$ receptor agonists with procognitive effects have been reported (Jensen et al., 2013). For further evidence, see also Meneses (2002) and Puig and Gulledge (2011).

According to Hanks and Gonzalez-Maeso (2013), psychedelic 5-HT$_{2A}$ receptor agonists LSD and DOI, but not lisuride, enhance trace conditioning of the nictitating membrane response in rabbits (a simple associative learning of a motor response), an effect reversed by 5-HT$_{2A/2C}$ receptor antagonists. Fear memory in a trace conditioning paradigm was also affected by activation of the 5-HT$_{2A}$ receptor in rats, and posttraining administration of the 5-HT$_{2A}$ receptor agonist (4-bromo-3,6-dimethoxybenzocyclobuten-1-yl)methylamine hydrobromide (TCB-2)115 enhanced subsequent freezing in a trace fear conditioning test (Hanks and Gonzalez-Maeso, 2013).

Although the role of 5-HT$_{2A/2B/2C}$ receptors in memory is unclear, the above and other studies suggest new horizons. For instance, evidence reported by Blasi et al. (2013) suggests that HTR2A affects 5-HT$_{2A}$ receptor expression and functionally contributes to genetic modulation of known endophenotypes of schizophrenia-like higher level cognitive behaviors and related prefrontal activity, as well as response to treatment with olanzapine. Moreover, true but not false memories seem to be associated with the HTR2A gene (Zhu et al., 2013).

CHAPTER 8

5-HT$_3$ Receptor

The 5-HT$_3$ receptor has a long date place in learning and memory investigation (Costall, 1993; Walstaba et al., 2010; Zhang et al., 2012). It should be noted that many 5-HT$_3$ receptor antagonists are licensed for the treatment of chemotherapy-induced (CINV), radiotherapy-induced, and postoperative nausea and vomiting (Thompson, 2013). In humans, there are five 5-HT$_3$ receptor subunits (5-HT$_{3A-E}$) (Hothersall et al., 2013; Barnes et al., 2009) with homomeric 5-HT$_{3A}$ and hetero-meric 5-HT$_{3AB}$ receptors being the most common and best character-ized. The 5-HT$_3$ receptors are found in many regions of the CNS including the hippocampus, EC, and frontal cortex and also play an important role in the enteric nervous system (Hothersall et al., 2013).

In their seminal work, Buhot et al. (2003c) highlighted that converg-ing evidence suggests that the administration of 5-HT$_{2A/2C}$ or 5-HT$_4$ receptor agonists or 5-HT$_{1A}$ or 5-HT$_3$ and 5-HT$_{1B}$ receptor antago-nists prevents memory impairment and facilitates learning in situations involving a high cognitive demand. In contrast, antagonists for 5-HT$_{2A/2C}$ and 5-HT$_4$, or agonists for 5-HT$_{1A}$ or 5-HT$_3$ and 5-HT$_{1B}$ generally have opposite effects. A better understanding of the role played by these and other serotonin receptor subtypes in learning and memory is likely to result from the recent availability of highly specific ligands, such as 5-HT$_{1A}$, 5-HT$_{1B}$, 5-HT$_{2A-2C}$ (as well as 5-HT$_{3-7}$) receptor antagonists, and new molecular tools, such as gene KO mice, especially inducible mice in which a specific genetic alteration can be restricted both temporally and anatomically (Buhot et al., 2003c).

Aiming to investigate further the role of 5-HT$_3$ receptor involved in learning and memory (Hong and Meneses, 1996), the specific 5-HT$_3$ receptor agonist 1-(m-chlorophenyl)-biguanide (mCPBG) and the 5-HT$_3$ receptor antagonists ondansetron and tropisetron were tested in the Pavlovian/instrumental autoshaping task. mCPBG impaired reten-tion of the CR, whereas tropisetron and ondansetron improved it; in other animals, the serotonergic depleter p-chloroamphetamine (PCA) alone did not affect CR but was able to block the effects of the 5-HT_3 ligands. These data suggest that the actions of 5-HT_3 compounds could be due to their interaction with presynaptic 5-HT_3 receptors (Hong and Meneses, 1996). Another 5-HT_3 receptor antagonist tropisetron enhances novel object recognition (NOR) in intact female rats (Sawyer et al., 2012) by improving the recognition of familiar information. It should be noted that in the last few years, a growing number of publications appeared related to 5-HT_3 receptor.

For instance, Rajan et al. (2011) reported that *Bacopa monniera* is a well-known medhya-rasayana (memory enhancing and rejuvenating) plant in Indian traditional medical system of Ayurveda. The effect of a standardized extract of *B. monniera* (BESEB CDRI-08) on serotonergic receptors and its influence on other neurotransmitters during hippocampal-dependent learning were evaluated. Wistar rat pups receiving a single dose of BESEB CDRI-08 during postnatal days 15−29 showed higher latency during hippocampal-dependent learning accompanied with enhanced 5HT_{3A} receptor expression, serotonin, and acetylcholine levels in hippocampus (Rajan et al., 2011). Furthermore, 5HT_{3A} receptor agonist mCPBG impaired learning in the passive avoidance task followed by reduction of 5HT_{3A} receptor expression, 5-HT, and acetylcholine levels. Administration of BESEB CDRI-08 along with mCPBG attenuated mCPBG-induced behavioral, molecular, and neurochemical alterations (Rajan et al., 2011). These authors suggest that BESEB CDRI-08 possibly acts on the serotonergic system, which in turn influences the cholinergic system through 5-HT_3 receptor to improve the hippocampal-dependent task (Rajan et al., 2011). In addition, Preethi et al. (2012) reported the participation of microRNA 124-CREB pathway in a parallel memory enhancing mechanism. This evidence is providing, the basis for better understanding functioning of muscle-type and neuronal nicotinic acetylcholine receptors (nAChRs), as well as of other Cys-loop receptors, including 5-HT_3, glycine, $GABA_A$, and some other (Tsetlin et al., 2009).

Notably, Boess et al. (2013) have reported that agonists of $\alpha7$ nAChRs may have therapeutic potential for the treatment of cognitive deficits. Their study describes the *in vitro* pharmacology of the novel $\alpha7$ nAChR agonist/serotonin 5-HT_3 receptor, concluding it improved performance in several learning and memory tests in both rats and mice, supporting the hypothesis that $\alpha7$ nAChR agonists may provide a novel therapeutic strategy for the treatment of cognitive deficits in AD or schizophrenia.

In addition, Park and Williams (2012) highlight that the changes in 5-HT$_3$ receptor activity influence memory and emotional regulation, the two essential components underlying the successful extinction of conditioned fear. They determined that blocking 5-HT$_3$ receptors with granisetron influences the extinction of fear, concluded that their studies reveal the beneficial effects of 5-HT$_3$ receptor activity in improving new learning associated with extinction of fearful memories, and suggest that these actions could be mediated through influences on central GABAergic systems (Park and Williams, 2012).

Barnes et al. (2009) highlight that the 5-HT$_3$ receptor is a cation-selective ion channel of the Cys-loop superfamily and its activation in the central and peripheral nervous systems evokes neuronal excitation and neurotransmitter release. Barnes et al. (2009) revise the relation-ship between the structure and function of the 5-HT$_3$ receptor. 5-HT$_{3A}$ and 5-HT$_{3B}$ subunits are well-established components of 5-HT$_3$ recep-tors but additional HTR3C, HTR3D, and HTR3E genes expand the potential for molecular diversity within the family. Importantly, according to Barnes et al. (2009) studies upon the relationship between subunit structure and the ionic selectivity and single channel conduc-tances of 5-HT$_3$ receptor have identified a novel domain (the intracel-lular MA stretch) that contributes to ion permeation and selectivity. Conventional and unnatural amino acid mutagenesis of the extracellu-lar domain of the receptor has revealed residues within the principle (A-C) and complementary (D-F) loops, which are crucial to ligand binding (Barnes et al., 2009). An area requiring much further investiga-tion is the subunit composition of 5-HT$_3$ receptors that are endogenous to neurons and their regional expression within the CNS. Barnes et al. (2009) conclude by describing recent studies that have identified numerous HTR3A and HTR3B gene polymorphisms that impact upon 5-HT$_3$ receptor function, or expression, and consider their relevance to (patho)physiology. Likewise, Newman et al. (2013) identified Cl-indole as a relatively potent and selective PAM of the 5-HT$_3$ receptor; such compounds will aid investigation of the molecular basis for allosteric modulation of the 5-HT$_3$ receptor and may assist the discovery of novel therapeutic drugs targeting this receptor. Importantly, Kondo et al. (2013) reported that analysis of 5-HT$_{3A}$ receptor KO mice in fear conditioning paradigms revealed that this is not required for the acquisition or retention of fear memory but is essential for the extinction of contextual and tone-cued; suggesting that the 5-HT3A receptor could be a key molecule regulating fear memory processes and a potential therapeutic target for fear disorders.

CHAPTER 9

5-HT$_4$ Receptor

5-HT$_4$ receptor possesses a long tradition in investigation (Bockaert et al., 2011), particularly, its role in learning and memory had been investigated for diverse groups (Buhot et al., 2003c; King et al., 2008; Maillet et al., 2004; Marchetti et al., 2011; Roman and Marchetti, 1998; Shen et al., 2011). In general terms, 5-HT$_4$ receptor agonists seem to facilitate memory in diverse memory tasks, including passive avoidance, autoshaping, and olfactory associative discrimination. In an attempt to further clarify 5-HT$_4$ receptors' role in memory, the expression of this receptor in passive (P3) untrained and autoshaping (A3) trained (3 sessions) adult (3 months) and old (P9 or A9; 9 months) male rats was determined by autoradiography (Manuel-Apolinar et al., 2005). Adult trained (A3) rats showed a better memory respect to old trained (A9). Using [H-3]GR113808 as ligand for 5-HT$_4$ receptor expression, 29 brain areas were analyzed, 16 areas of A3 and 17 of A9 animals displayed significant changes. For instance, the medial mammillary nucleus of A3 group showed diminished 5-HT$_4$ receptor expression, and in other 15 brain areas of A3 or 10 of A9 animals, 5-HT$_4$ receptor was increased. Thus, for A3 rats, 5-HT$_4$ receptor was augmented in olfactory lobule, caudate putamen, fundus striatum, hippocampal CA2, retrosplenial, frontal, temporal, occipital, and cingulate cortices. Also, 5-HT$_4$ receptors were increased in olfactory tubercule, hippocampal CA1, parietal, piriform, and cingulate cortices of A9. However, hippocampal CA2 and CA3 areas, and frontal, parietal, and temporal cortices of A9 rats expressed less 5-HT$_4$ receptor. These findings suggest that serotonergic activity, via 5-HT$_4$ receptor in hippocampal, striatum, and cortical areas, is associated with memory function and provides further evidence for a complex and regionally specific regulation over 5-HT receptor expression during memory formation (Manuel-Apolinar et al., 2005).

The above data suggest that 9-month-old rats showing poor memory retention relative to adult rats displayed diminished expression of 5-HT$_4$ receptor in CA2 and CA3 hippocampal areas and temporal cortex. Although this latter brain area showed a nonsignificant diminished expression of 5-HT$_4$ receptors in AD brains relative to its control age group in the Reynolds et al. study (Manuel-Apolinar et al., 2005), notwithstanding there is an interesting parallel result in frontal cortex (area 11) of AD patients and old rats; in both groups expression of 5-HT$_4$ receptor was decreased. Notably, Pavlovian autoshaping seems to be associated with an increase in 5-HT content in the PFC in rats, likely due to enhance enzymatic activity of tryptophan hydroxylase increase, which would increase synthesis of 5-HT (Tomie et al., 2003, 2004). Notably, it is well known that serotonergic activity depends on tryptophan availability, the rate-limiting enzyme tryptophan hydroxylase, monoamine oxidase, re-uptake sites, and 5-HT receptors, which also might be influenced by learning, memory, and aging (Manuel-Apolinar et al., 2005; Tellez et al., 2010, 2012b). Together, these data are in line with the notion of contribution of 5-HT$_4$ receptors in the physiology, pathophysiology, and therapeutic of learning and memory (Perez-Garcia and Meneses, 2008a).

Further support for the above conclusion comes from a recent work in which we aimed to determine, by autoradiography and using [^{3}H] 5-HT as ligand, 5-HT receptor expression in passive (untrained) and autoshaping trained (3 sessions) adult (3 months) and old (9 months) male rats (Perez-Garcia and Meneses, 2008a). Trained adult rats presented better retention than old animals. RN of adult and old trained rats expressed less receptors on medial and dorsal, respectively, and hippocampal CA1 area and dentate gyrus of adult trained rats expressed less 5-HT receptors, while dentate gyrus of old animals increased them. Basomedial amygdaloid nucleus in old trained rats expressed more 5-HT receptors, while in the basolateral amygdaloid nucleus, they were augmented in both groups. Autoshaping training decreased or did not change 5-HT receptors expression in caudate putamen of adult or old animals. This profile of 5-HT receptor expression and localization is consistent with previous reports and suggests that memory formation and age may modulate 5-HT receptor expression in brain areas relevant to memory systems. Moreover, it should be noticed that recently Campan et al. (2004) reported that 5-HT$_4$ receptor-null mice are less reactive in novel environments. The 5-HT$_4$ receptor-null mice were placed in an open field and their locomotion was monitored for 30 min and the test was repeated for three consecutive days to evaluate habituation to a novel environment, which could also represent a simple learning and memory task. The mutant mice displayed an overall decrease in the traveled path length only during the first day of exposure, as compared with WT animals in both the periphery and center of the open field (Campan et al., 2004). According to them, in contrast, there was no difference in locomotion between the mutant and WT animals in their home cages and less reactivity in 5-HT$_4$ receptor-null mice has also been found in the elevated plus maze and alley tests. In addition, the mutant mice did not exhibit impairment in the rotarod test, which demonstrates that there is no deficit in motor coordination in 5-HT$_4$ receptor-null mice. In these mice, pentylenetetrazol-induced convulsive responses were enhanced, suggesting an increase in neuronal network excitability. Campan et al. (2004) conclusion was that these results suggest that the 5-HT$_4$ receptor-null mice display decreased reactivity to novelty rather than locomotor impairment. But could these same data be also suggesting an improved habituation? (Manuel-Apolinar et al., 2005). Kemp and Manahan-Vaughan (2005) reported that 5-HT$_4$ receptor exhibits frequency-dependent properties in synaptic plasticity and behavioral metaplasticity in the hippocampal CA1 region *in vivo* finding that key role for 5-HT$_4$ receptors in the regulation of synaptic plasticity and the determination of the particular properties of stored synaptic information.

Moreover, apparently, the absence of 5-HT$_4$ receptors reduced the time of habituation during the first 15 min of observation. Of course, though more information on other learning and memory tasks is necessary, wide pharmacological evidence indicates that while 5-HT$_4$ receptor agonists facilitate learning and memory, antagonists for these receptors have no effect or may impair performance (Manuel-Apolinar et al., 2005). For instance, in an olfactory associative discrimination task (Lamirault and Simon, 2001), blockade of 5-HT$_4$ receptor before the third training session induced a consistent deficit in associative memory (place and object recognition in young adult rats) during the following training sessions, and this deficit was absent when the antagonist was injected together with selective 5-HT$_4$ receptor agonists. Moreover, in models of memory deficit, partial 5-HT$_4$ receptor agonists were used to reverse WM (spontaneous alternation) deficits induced by scopolamine, which further confirm the therapeutic potential of such ligands in the treatment of cognitive alterations associated with short-term WM disorders and cholinergic hypofunction (Lelong et al., 2006). Also relevant is the finding that pretraining administration of the 5-HT$_4$ receptor agonists enhanced the acquisition of autoshaping response, while posttraining administration of either of these agents impaired memory consolidation (Meneses and Hong, 1997). In addition, when 5-HT$_4$ receptor was stimulated in pretreated rats with 5-HT$_4$ receptor antagonists, the decrement induced by both agonists was reversed. Interestingly, Lamirault and Simon (2001) found an increased memory consolidation of object recognition in old rats with a 5-HT$_4$ receptor agonist. Importantly, RS67333 (partial agonist for 5-HT$_4$ receptors) facilitated both autoshaping STM (1.5 h) and LTM (24 and 48 h) protocols (Meneses, 2007b), but its pretraining administration improved memory acquisition of autoshaping response, while posttraining had no effect on STM (24 h) and LTM (48 h) protocols (Manuel-Apolinar et al., 2005), thus highlighting the importance of timing of drug administration (Meneses, 2013).

Certainly, discrepancies among the above works might be related to age of rats, drugs, and nature of behavioral tasks and protocols of training/testing used (Meneses, 2013). Indeed, inasmuch as memory formation studies have identified memory phases differing on relative contribution basis of brain areas, neurotransmitters, receptors, signal transduction pathways, protein synthesis, and genes induction (Manuel-Apolinar et al., 2005; Perez-Garcia and Meneses, 2008a; Meneses, 2013; 2014). Actually, multiple 5-HT receptors have been identified in brain areas relevant to cognitive processes, and, possibly during memory formation and amnesia the expression of some of these receptors may have task- and regional-dependent variations (Meneses, 2013). Notably, receptors (Schiller et al., 2003) and transduction signals (Manuel-Apolinar and Meneses, 2004) are dynamic, time-dependent patterns of learning- and memory-induced regulatory processes.

Bockaert et al. (2011) highlight that 5-HT$_4$ receptor controls brain physiological functions, including learning and memory, feeding and mood behavior as well as gastrointestinal transit. 5-HT$_4$ receptor is one of the 5-HT receptors for which the available drugs and signaling knowledge are the most advanced (Bockaert et al., 2011). Several therapeutic 5-HT$_4$ receptor drugs have been commercialized; therefore, the hope that 5-HT$_4$ receptor could also be the target for brain diseases is reasonable, including AD, feeding-associated diseases such as anorexia and major depressive disorders (MDDs) (Bockaert et al., 2011; King et al., 2008; Shen et al., 2011). Notably, Cochet et al. (2013) found that 5-HT$_4$ receptor constitutively promote the nonamyloidogenic pathway of APP cleavage and interacts with ADAM10. In addition,

Haahr et al. (2012) found that the 5-HT$_4$ receptor seems to be associated with memory functions in the human hippocampus suggesting potentially pharmacological stimulation of the receptor may improve episodic memory. Finally, Eriksson et al. (2012c) reported that Flinders sensitive line (FSL) rats, a genetic rat model of depression, display a pronounced impairment of emotional memory function in the passive avoidance task, accompanied by reduced transcription of Arc in PFC and hippocampus; at the cellular level, FSL rats have selective reductions in levels of NMDA receptor subunits, 5-HT$_{1A}$ receptor, and MEK activity. Chronic treatment with escitalopram, but not with an antidepressant regimen of nortriptyline, restored memory performance and increased Arc transcription in FSL rats and procognitive effects could also be achieved by either disinhibition of 5-HT$_{1A}$ receptor/MEK/Arc or stimulation of 5-HT$_4$ receptor/MEK/Arc signaling cascades (Eriksson et al., 2012c). An interesting differential role of the basolateral amygdala 5-HT$_3$ and 5-HT$_4$ receptors about anxiolytic-like behaviors and emotional memory deficit in mice was identified (Chegini et al., 2013).

CHAPTER 10

5-HT$_5$ Receptor

Volk et al. (2010) reported that although the 5-HT$_5$ receptor subfamily was discovered more than 15 years ago, it is unambiguously the least known 5-HT receptor subtype. The G_i/G_0-mediated signal transduction and its intensive presence in raphe and other brainstem and pons nuclei suggest mechanisms similar to those of 5-HT$_1$ receptors, the ligands of which are already applied in the treatment of anxiety and migraine (Volk et al., 2010). Certainly, high concentrations of 5-HT$_5$ receptor in other key regions, including locus coeruleus, nucleus of the solitary tract, arcuate and suprachiasmatic nuclei of the hypothalamus indicate a wide range of physiological effects, thus its ligands are potential drug candidates in various areas, e.g., anxiety, sleep, incontinence, food intake, learning and memory, pain, or chemoreception pathways. This evidence has motivated several institutes and pharmaceutical companies to participate in the research of this field; however, despite extensive research, no selective agonist and only two selective antagonists have been identified until now (Volk et al., 2010). Importantly, Volk et al. (2010) provide an overview about 5-HT$_{5A}$ receptor ligands, structure, function, distribution, genetics, and possible therapeutic applications.

In a recent attempt to explore the role of 5-HT$_5$ receptor, Gonzalez et al. (2013) reported that posttraining injection of the selective 5-HT$_{5A}$ receptor antagonist SB-699551 at 0.3 mg/kg decreased performance during Pavlovian/instrumental autoshaping STM (1.5 h) and LTM (24 h) or at 1.0 and 3.0 mg/kg impaired LTM (24 h) relative to the vehicle animals. The serotonin precursor L-tryptophan (50.0 mg/kg) facilitated performance and in the interaction experiments, it attenuated the impairment effect induced by SB-699551 (either 0.1 or 3.0 mg/kg). Why did these doses of SB-699551 or L-tryptophan selectively alter specific phases of memory? We should bear in mind that many agents have "inverted U" dose−response curves in behavioral and mechanistic procedures of memory tasks, thus biphasic dose−response curves imply a "set point" for optimal (or worst) performance, such that under- or overactivation of the drug target has a deleterious (or improvement) effect (Millan et al., 2012). In addition, pharmacological manipulation of 5-HT systems (Meneses, 2007a) evidenced serial and parallel function between STM and LTM. For instance, the "5-HT tone via 5-HT_{1B} receptors" function in a serial manner from STM to LTM, whereas working in parallel using 5-HT_{1A}, 5-HT_{2A}, $5\text{-HT}_{2B/2C}$, 5-HT_4, or 5-HT_6 receptors (Meneses, 2007b). Hence, we might assume that the function of 5-HT_{5A} receptor (as antagonized by SB-699551) during STM and LTM is parallel and serial, which was dependent of the dose, while L-tryptophan (at 50.0 mg/kg) allowed STM and LTM (24 h) functioned in serial manner. Thus, L-tryptophan attenuated the SB-699551-induced impairment effect in both STM and LTM. Of course, future studies should clarify issues like why higher doses of L-tryptophan had no effects? Part of the answer might be the above mentioned by Millan et al. (2012). Moreover, systematic studies of other drugs stimulating 5-HT receptors revealed that except for DOI (agonist for $5\text{-HT}_{2A/2C}$ receptors) dose-dependently impaired STM and, at 10.0 mg/kg only impaired LTM; other 5-HT receptor agonists did not (e.g., dual $5\text{-HT}_{1A/7}$ receptor, 5-HT_{1B}, 5-HT_3, 5-HT_4, and 5-HT_7) (Meneses, 2007a).

It should be noted inasmuch as SB-699551 at the dose of 0.1 or 3 mg/kg has no effect on LTM (48 h), then how do we explain the decrease of the beneficial effect of L-tryptophan (50 mg/kg) on LTM memory (48 h) when it was administered with SB-699551 at the dose of 3.0 mg/kg? An alternative is that at this time L-tryptophan dose might be stimulating not only 5-HT_{5A} receptor but also others whose stimulation impaired LTM (e.g., 5-HT_3 receptor) (Meneses, 2007a). It is also not possible to exclude the interaction of other neurotransmission systems and convergent cell signaling.

How do we explain the few studies evaluating the involvement of 5-HT_5 receptor, even though selective antagonists for this receptor have been available for about a decade? Volk et al. (2010) suggested that since the 5-HT_{5B} receptor is not expressed in humans, no serious efforts have been undertaken to develop selective ligands for this receptor subtype. In addition, it should be noted that serotonergic systems in memory investigation remained little explored in the first years of the 1990s, in spite of availability of multiple pharmacological tools; certainly, at present growing number of papers are being published. For instance, 5-HT_6 and 5-HT_7 receptors represent a growing scientific interest (see below chapter 11).

Hence, it seems timely to highlight that the above data are consistent with the notion that 5-HT$_{5A}$ receptor might have a role in memory (Volk et al., 2010). 5-HT$_5$ receptor might play an inhibitory role (Goodfellow et al., 2012; Kassai et al., 2012). Inasmuch as genetic deletion of the inhibitory 5-HT$_{5A}$ receptor results in an unexpected, large increase in the inhibitory 5-HT$_{1A}$ receptor currents; according to Goodfellow et al. (2012) suggesting that the presence of functional prefrontal 5-HT$_{5A}$ receptor in normal rodents along with compensatory plasticity in 5-HT$_{5A}$ receptor KO mice testifies to the significance of this receptor in the healthy PFC. Of course, in the study by Gonzalez et al. (2013), systemic administration was used and considering that PFC, among other brain areas, is important for memory function and forgetting (Tellez et al., 2012a,b). Hence, intracerebral administration of SB-699551 might reveal important insights about the function of 5-HT$_{5A}$ receptor for memory and other functions and molecular mechanisms; certainly, the finding that SB-699551-L-tryptophan combination revealed the functional importance of this receptor (Gonzalez et al., 2013).

Thus, with the classification of multiple 5-HT receptors (Hannon and Hoyer, 2008), it has become clear that the stimulation or blockade of diverse 5-HT receptors modulates memory formation (Terry et al., 2008, 2011). It should be noted that Curtin et al. (2013) recently reported that 5-HT$_{5A}$ regulates cell excitability by modulation of a membrane conductance in the gold fish, which in turn influences the magnitude of sensorimotor gating and prepulse inhibition. In consequence, it is of no surprise that the 5-HT$_5$ receptor blockade affected autoshaping STM and LTM (24 h). As already mentioned, 5-HT$_5$ receptor occurs in hippocampus, cerebral cortex, amyloid nuclei, and RN (Kassai et al., 2012; Thomas, 2006) brain areas involved in learning and memory processes (Hampel et al., 2011; Molinuevo et al., 2012). This information provides further support to the contention that: (i) 5-HT pathways and receptors show a regional distribution in brain areas implicated in learning and memory; (ii) significant changes occur in brain 5-HT system functions as a result of AD, aging, memory formation, amnesia and antiamnesic effects (for review, see Lesch and Waider, 2012; Meneses, 2013; Meneses et al., 2011a,b; Rodríguez et al., 2012); (iii) 5-HT compounds modulate memory formation; in the present context, SB-699551 impaired STM and LTM. Inasmuch as this compound displays affinity for 5-HT$_{5A}$ receptor (Thomas et al., 2006), blockade of this receptor impaired these cognitive processes and as L-tryptophan attenuated this SB-699551 effect; hence, a serotonergic mechanism is implicated and an interesting and heuristic possibility is the investigation of 5-HT$_{5A}$ receptor agonists (Gonzalez et al., 2013).

Concluding, the role of 5-HT$_{5A}$ receptor in cognitive processes might be critical. Of course, it remains to be determined by many other aspects; for instance, further studies with selective 5-HT$_5$ receptor agonists and other antagonists and using other behavioral tasks are needed (Gonzalez et al., 2013).

CHAPTER 11

5-HT$_6$ Receptor

5-HT$_6$ receptor possesses an emergent and growing tradition in learning and memory investigation, including patents reviews (Ivachtchenko and Ivanenkov, 2012; Ivachtchenko et al., 2012; Ruiz and Oranias, 2010), and several important reviews have been published in the last few years (Fone, 2006; Geldenhuys and Van der Schyf, 2009; Gong et al., 2012; Holenz et al., 2006; Johnson et al., 2008; King et al., 2008; Liu and Robichaud, 2009; Marin et al., 2012a,b; Mitchell and Neumaier, 2005; Ramírez, 2013; Rossé and Schaffhauser, 2010; Russel and Dias, 2002; Terry et al., 2008; Upton et al., 2008; Wilson and Terry, 2009; Woolley et al., 2003). Notably, some 5-HT$_6$ receptor drugs represent potential antidementia treatment (Johnson et al., 2008; Geldenhuys and Van der Schyf, 2011; King et al., 2008; Maher-Edwards et al., 2009; Meneses, 2013; Rossé and Schaffhauser, 2010; Terry et al., 2008; Upton et al., 2008; Yun and Rhim, 2011a,b; Witty et al., 2009). Pharmacological blockade of 5-HT$_6$ receptor had been shown to produce promnesic or antiamnesic effects (or both) in a number of memory tasks, including water maze, passive avoidance, autoshaping, fear conditioning, novel object recognition, or social memory (Ramirez, 2013). Likewise, emergent evidence indicates that 5-HT$_6$ receptor agonists seem to facilitate memory in diverse memory tasks, including novel object recognition and autoshaping (King et al., 2008; Meneses et al., 2011a,b; Ramirez, 2013). The reasons for apparent paradox of promnesic and/or antiamnesic effects of 5-HT$_6$ receptor agonists and antagonists are unclear (see e.g., Woods et al., 2013; Meneses, 2013).

The 5-HT$_6$ receptor offers an excellent timely example; when comparing 5-HT$_6$ receptor drugs and/or protocols of training, it is not surprising to observe some convergent and divergent findings between appetitive associative Pavlovian/instrumental autoshaping versus water maze (spatial memory) or aversive paradigms such as inhibitory avoidance or other memory behavioral tasks (Meneses et al., 2011b). Certainly, the development of potent and selective 5-HT$_6$ receptor antagonists has been crucial in the clarification of the role of 5-HT$_6$ receptor role on memory; in addition, the findings that 5-HT$_6$ receptor agonists can improve memory are providing new insights. Likewise, the study of regulation by drugs, age, and/or memory themselves on 5-HT$_6$ receptor expression and/or signal cascades is very important. Clarifications also about whether 5-HT$_6$ receptor drugs are acting as inverse agonists/antagonists will be very crucial (see below). Likewise, if the memory tasks modify the expression of 5-HT neurobiological markers (e.g., receptors, SERT, signaling), will be crucial to determine (Marcos et al., 2010; Tellez et al., 2012a,b). These aspects are summarized and contextualized below.

11.1 NEURAL MARKERS AND MEMORY TASKS

The above considerations are quietly important in the context of neural markers and memory tasks (see above the respective section about memory tasks). For instance, rats overexpressing 5-HT$_6$ receptor in dorsomedial striatum, but not dorsocentral striatum, showed impaired performance in a simple operant learning task (a striatum-dependent learning model), but not in the hippocampus-dependent water maze task (Mitchell et al., 2007). This impairment effect was appreciable at third instrumental testing session or the second extinction session on performance of previously acquired instrumental conditioning (Mitchell et al., 2007), suggesting a specific time window.

It should be noted that in Pavlovian/instrumental autoshaping task, memory formation involves a serotonergic tone (at least) via 5-HT$_6$ receptor protein and mRNA, suppressing its expression and its pharmacological blockade improves memory formation (Meneses et al., 2004b). Under amnesic conditions, 5-HT$_6$ receptor (either protein or mRNA) is completely suppressed or slightly reduced (depending on the pharmacological model of amnesia); in contrast, the 5-HT$_6$ receptor antagonist SB-399885 improved memory or amnesia and reversed the expression of 5-HT$_6$ receptor, which was increased or reestablished, respectively (Huerta-Rivas et al., 2010; Meneses et al., 2004b). Moreover, memory formation on the water maze (MWM) downregulated 5-HT$_6$ receptor protein and mRNA receptor expression and the administration of the selective 5-HT$_6$ receptor antagonist SB-271046 induced an increase in pCREB1 levels while CREB2 levels were significantly reduced (Marcos et al., 2010). However, although SB-271046 was able to improve retention in the MWM, no further changes in pCREB1 or CREB2 levels were observed, but the MWM alone significantly increased pERK1/2 levels and further increases were seen with SB-271046 during the MWM. Certainly, Ly et al. (2013) had recently noted that the 5-HT$_6$ antagonist SAM-531 failed to potentiate theta power, which is characteristic (according to these authors) of many procognitive substances, indicating that 5-HT$_6$ receptor might not tonically modulate hippocampal oscillations and sleep—wake patterns.

In addition, SB-271046 reverses scopolamine-disrupted consolidation of a passive avoidance task and ameliorates spatial task deficits in aged rats (Foley et al., 2004). Likewise, Da Silva Costa-Aze et al. (2012) tested SB-271046, alone or in combination with scopolamine, on WM (spontaneous alternation task in the T-maze), recognition memory (place recognition), and aversive learning (passive avoidance), reporting that SB-271046 alone failed to affect WM, recognition memory, and aversive learning performances. In contrast, SB-271046 was able to reverse the scopolamine-induced deficits in WM and those of acquisition and retrieval of aversive learning (dose-dependent effect); scopolamine-induced deficits in episodic-like memory (acquisition and retrieval) were partially counteracted by 5-HT$_6$ receptor blockade (Da Silva Costa-Aze et al., 2012). According to them, the modulation between 5-HT$_6$ receptor and the cholinergic system appears to be predominant for WM and aversive learning, but not for other types of memory (i.e., episodic-like memory). Interactions between 5-HT$_6$ receptor and alternative neurotransmission systems (i.e., glutamatergic system) should be further studied (Da Silva Costa-Aze et al., 2012); the respective involvement of these interactions in the memory disorders related to aging and neurodegenerative diseases is of pivotal importance regarding the possible use of 5-HT$_6$ receptor antagonists in the treatment of memory disorders in humans (Da Silva Costa-Aze et al., 2012). In the water maze, administration of SB-271046-A or SB-357134-A had no effect on learning *per se*. However, both compounds produced a significant improvement in retention of a previously learned platform position when tested 7 days after training. By contrast, the acetylcholinesterase inhibitor, donepezil had no effect in this task (Rogers and Hagan, 2001).

Callaghan et al. (2012) examined the effects of acute and chronic administration of SB742457 on performance in a delayed non-matching-to-sample task (DNMS), which was used to identify neurocognitive differences between middle-aged (MA, 13 months) and young adult

(YG, 3 months) rats. Callaghan et al. (2012) found that MA rats have significantly lower performance in the DNMS task compared to YG rats. Acute administration of SB742457 significantly improved performance of the MA rats and its chronic administration reversed the age-related deficit of the MA to match their performance to that of YG rats; these improvements were observed for 1 week post-SB742457 treatment cessation. The acute and chronic effects of this treatment suggest that there is both an immediate effect on neurotransmitter action and potentially a longer term modification of synaptic plasticity; indicating a role for modulation of the serotonergic system in the development of cognition-enhancing agents (Callaghan et al., 2012).

In addition, 5-HT$_6$ receptor agonists might be useful for their pro-memory and/or antiamnesic effects (Woods et al., 2012) acting on the cholinergic and/or glutamatergic neurons or alternatively, 5-HT$_6$ receptor antagonists and agonists might operate via modulation of distinct intracellular signaling pathways (Romero et al., 2006; Woods et al., 2012). As already noted (see above) density and 5-HT$_6$ levels were significantly decreased in a cohort of AD patients (Marcos et al., 2008). For other serotonergic neurobiological markers showing changes related to AD, and/or aging as well as other diseases with dysfunctional memory (see other parts of the present paper). Importantly, Nikiforuk et al. (2013) reported that EMD 386088 ameliorates ketamine-induced deficits in attentional set shifting and novel object recognition, but not in the prepulse inhibition in rats.

Certainly, Gravius et al. (2011), Linder et al. (2003), Russel and Dias (2002) did not find evidence of antiamnesic or promnesic effects of 5-HT$_6$ receptor antagonists. The reasons for these discrepant results had been not discussed, but probably methodology differences exist (Perez-Garcia and Meneses, 2005a; Gonzalez et al., 2013; Woods et al., 2012), including drugs, doses, and memory tasks used, devices for measuring memory and protocols of training/testing (Meneses, 2013; Meneses et al., 2011a). As already mentioned, development of more effective 5-HT$_6$ receptor drugs had been also an important advance.

CHAPTER 12

5-HT$_7$ Receptor

It should be noted that the 5-HT$_7$ receptor is a more recently discovered GPCR for serotonin, and the functions and possible clinical relevance of this receptor are not yet fully understood (Roberts and Hedlund, 2012). According to them, animal models of learning and memory and other techniques have implicated the 5-HT$_7$ receptor in such processes by using a combination of pharmacological and genetic tools targeting the receptor to evaluate effects on behavior and cellular mechanisms (Roberts and Hedlund, 2012). In behavioral tests such as the Barnes maze, contextual fear conditioning and novel location recognition that involve spatial learning and memory considerable evidence supports an involvement of the 5-HT$_7$ receptor, including studies of mRNA expression and cellular signaling as well as in electrophysiological experiments (Roberts and Hedlund, 2012). Especially interesting are the subtle but distinct effects observed in hippocampus-dependent models of place learning where impairments have been described in mice lacking the 5-HT$_7$ receptor or after administration of a selective antagonist (Roberts and Hedlund, 2012). While more work is required, it appears that 5-HT$_7$ receptor is particularly important in allocentric representation processes and in instrumental learning tasks both procognitive effects and impairments in memory have been observed using pharmacological tools targeting the 5-HT$_7$ receptor (Roberts and Hedlund, 2012). Roberts and Hedlund (2012) conclude that the use of pharmacological and genetic tools in animal studies of learning and memory suggests a potentially important role for the 5-HT$_7$ receptor in cognitive processes (Cifariello et al., 2008; Leopoldo et al., 2010). Notably, Leopoldo et al. (2011) provide recent advances targeting 5-HT$_7$ receptor in the context of structure—activity relationships and potential therapeutic applications in CNS disorders.

Considering that 5-HT$_{1A}$ and 5-HT$_7$ receptors seem to interact in pharmacological terms (see above), hence, their functional role in memory has been studied with specific antagonists (5-HT$_{1A}$ receptor, WAY100635) and (5-HT$_7$ receptor, SB-269970 or DR 4004), which decreased the improvement of performance in Pavlovian/instrumental autoshaping tasks, produced by the dual $5\text{-HT}_{1A/7}$ agonist 8-OH-DPAT to levels lower than control animals (Meneses, 2004). These same antagonists decreased the performance levels after PCA lesion of the 5-HT system, which has no effect on memory consolidation (Meneses, 2004), but SB-269970 or DR 4004 reversed amnesia induced by scopolamine and dizocilpine (Meneses, 2004). In addition, Gasbarri et al. (2008) reported that SB-269970 improved memory by decreasing the number of errors in test phase and, thus, affecting reference memory, without affecting WM; thus, 5-HT_7 receptor blockade had procognitive effect, when the learning task implicated a high degree of difficulty (Gasbarri et al., 2008; Nikiforuk and Popik, 2013). The above data indicate a role for 5-HT_{1A} and/or 5-HT_7 receptors in memory formation, and are supporting the hypothesis that serotonergic, cholinergic, and glutamatergic systems interact in cognitively impaired animals (Meneses, 2004, 2013, 2014). Also 5-HT_7 receptor seems to be involved in the degree difficult of memory or cognitive demand (Gasbarri et al., 2008). Importantly, the above findings support a potential role for 5-HT_{1A} and 5-HT_7 receptors in the pathophysiology and/or treatment of cognitive deficits and the mechanism of action of atypical antipsychotic drugs (Meneses, 2013; Perez-Garcia and Meneses, 2005b) or memory disorders in other psychiatric diseases.

Supporting the above conclusions, the analysis of the time course $(0-120\ \text{h})$ of autoshaped responses revealed that in progressive memory, mRNA 5-HT_{1A} or 5-HT_7 receptors expression was monotonically augmented or declined in PFC, hippocampus, and RN, respectively (Perez-Garcia and Meneses, 2009). At $24-48\ \text{h}$ acutely 8-OH-DPAT enhanced memory and attenuated mRNA 5-HT_{1A} more than 5-HT_7 receptor expression respect to saline group. WAY100635 or SB 269970 alone had no effect on memory but partially blocked or completely reversed mRNA expression, respectively. WAY100635 plus SB-269970 impaired memory consolidation and suppressed 5-HT_{1A} or 5-HT_7 receptor expression. Thus, when both 5-HT_{1A} and 5-HT_7 receptors were stimulated by 8-OH-DPAT (subeffective low doses) under memory consolidation, subtle changes emerged, not evident at behavioral level though detectable at genes expression. Notably, high levels of efficient memory were maintained even when either 5-HT_{1A} or 5-HT_7 receptor was down- or upregulated (Perez-Garcia and Meneses, 2009). While these receptors might be not essential for memory, however, the pharmacological manipulation of these receptors would allow neuroplasticity margins. Moreover, the 5-HT$_7$ receptor agonist AS19 facilitated memory (LTM 24 and 48 h protocol) consolidation, decreased or increased hippocampal 5-HT$_7$ and 5-HT$_{1A}$ receptors expression (Perez-Garcia et al., 2006). Considering that serotonergic changes are prominent in AD patients with an earlier onset of disease the above approach might be useful in the identification of functional changes associated with memory formation, memory deficits, and reversing or even preventing these deficits.

Understanding the cellular and molecular mechanisms underlying the formation and maintenance of memories is a central goal of the neuroscience community (Zovkic et al., 2013). For them, the (structural and functional) plasticity is dependent on a well-regulated program of neurotransmitter release, postsynaptic receptor activation, intracellular signaling cascades, gene transcription, and subsequent protein synthesis (Zovkic et al., 2013), including cAMP (Kandel, 2001, 2012). The cAMP is a second messenger and a central component of intracellular signaling pathways that regulate a wide range of biological functions, including memory (Kandel, 2001). As, Izquierdo et al. (2006) found augmented cAMP production (immediately and 90 min later) during memory formation in passive avoidance task. Hence, the time course of memory formation was determined in an autoshaping learning task (Perez-Garcia and Meneses, 2008b), using the STM (1.5 h) and LTM (24 and 48 h) protocols and considering the data of untrained and treated animals as the basal values of cAMP production relative to those of autoshaping trained and treated animals (Perez-Garcia and Meneses, 2008b); memory formation (saline group) *per se* increased cAMP production relative to untrained animals. However, memory formation plus 8-OH-DPAT elicited lower values of cAMP in PFC, while in the untrained group cAMP was increased (Perez-Garcia and Meneses, 2008b). It should be noted that the values of cAMP in control trained and untrained animals ranged in the former between 0 and 2000 and in the latter between 0 and 500 (Perez-Garcia and Meneses, 2008b). Thus, cAMP was increased or decreased by the stimulation (8-OH-DPAT) of 5-HT$_{1A/7}$ receptor if memory was improving but the opposite occurred in the lack of memory (Perez-Garcia and Meneses, 2008b). Confirming the 5-HT$_7$ receptor participation, the 5-HT$_7$ receptor agonist AS19 improved memory and elicited higher (RN and PFC) cAMP production in both trained and untrained animals; importantly, AS19 intrahippocampal administration facilitated memory and produced even significantly higher cAMP scores in RN, hippocampus, and PFC (Perez-Garcia and Meneses, 2008b). The effects of 8-OH-DPAT and AS19 were partially or completely blocked by WAY100635 or SB-269970, respectively; thus, providing further support to the notion that 5-HT_{1A} and 5-HT_7 receptors were involved in these effects (Perez-Garcia and Meneses, 2008a,b). Certainly, it should be noted that different protocols of training/testing and memory (e.g., STM 1.5 h and LTM 24 and 48 h vs. LTM 24 and 48 h) seem to be associated differentially to 5-HT_{1A} and 5-HT_7 receptors expression and cAMP production plus memory formation and drug administration. How 5-HT_{1A} and 5-HT_7 receptors or AMPc are contributing in this regard?

Evidence (Perez-Garcia and Meneses, 2009) indicates that in Pavlovian/instrumental autoshaping the RT-PCR analysis of 5-HT_{1A} and 5-HT_7 receptor mRNA in RN, hippocampus, and PFC at 0, 1.5, 24, and 48 h following training/testing sessions revealed an initial (0 h) 5-HT_{1A} receptor low level followed by a sustained growing, being significant at 48 and 120 h (Perez-Garcia and Meneses, 2009). In contrast, 5-HT_7 receptor mRNA at 0 h was highest in the three brain areas and as training/testing sessions augmented its expression significantly, being the lowest level at 120 h; notably, hippocampal 5-HT_7 receptor mRNA expression showed the lowest values at 1.5 and 120 h, while at 24 h it showed a recovery and gradually declining onwards. Further analysis confirmed that as training/testing sessions were accumulated 5-HT_{1A} receptor expression augmented in the three brain areas; in contrast, for 5-HT_7 receptor the opposite pattern was evident, as training and testing sessions were accumulated, their expression declined (Perez-Garcia and Meneses, 2009). As there were significant increments in both the %CR and mRNA expression at 48 h relative to 1.5 and 24 h testing times, hence, 24 and 48 h had being selected for further memory, pharmacological and mRNA experimental analyses (Perez-Garcia and Meneses, 2009). The above evidence indicates 5-HT_{1A} and 5-HT_7 receptor levels of expression (Perez-Garcia and Meneses, 2009) as well as cAMP production (Perez-Garcia and Meneses, 2008b) depending on timing of memory formation. Also, certainly, the above data illustrate the importance and complexity of cAMP production, as well as 5-HT_{1A} or 5-HT_7 receptor expression, plus drug administration in the signaling of mammalian memory formation.

Importantly, Aubert et al. (2013) reported that repeated (16 weeks) administration of 8-OH-DPAT (low doses) affected the gene classes (using a marmoset-specific microarray) important to neural development mPFC, medial preoptic area (mPOA), and raphe dorsal nuclei; neuro-transmission (mPOA), energy production (mPFC and mPOA), learning and memory (hippocampal CA1), and intracellular signal transduction (dorsal raphe nuclei). Oxytocin (OXT) in the mPOA and SERT in the DRN were strongly increased by 8-OH-DPAT. 5-HT$_{1A}$ receptor tended to increase in the mPFC, while 5-HT$_7$ receptor was decreased in the CA1 (Aubert et al., 2013). Likewise, Eriksson et al. (2012a) found that with the elevation of 5-HT by the SSRI fluoxetine had no effect by itself, but facilitated emotional memory (passive avoidance task) performance when combined with the 5-HT$_{1A}$ receptor antagonist NAD-299. This facilitation was blocked by the selective 5-HT$_7$ receptor antagonist SB-269970, revealing excitatory effects of fluoxetine via 5-HT$_7$ receptor. The enhanced memory retention by NAD-299 was blocked by SB-269970, and according to Eriksson et al. (2012a), indicating that reduced activation of 5-HT$_{1A}$ receptor results in enhanced 5-HT stimulation of 5-HT$_7$ receptor. The putative 5-HT$_7$ receptor agonists LP-44 when administered systemically and AS19 when administered both systemically (AS19 5 or 10 mg/kg) and into the dorsal hippocampus (AS19 10 μM) failed to facilitate memory, and according to Eriksson et al. (2012a) this finding is consistent with the low efficacy of LP-44 and AS19 to stimulate protein phosphorylation of 5-HT$_7$ receptor activated signaling cascades. In contrast, increasing doses of the dual 5-HT$_{1A/7}$ receptor agonist 8-OH-DPAT impaired memory, while coadministration with NAD-299 facilitated emotional memory in a dose-dependent manner; this facilitation was blocked by SB-269970 indicating 5-HT$_7$ receptor activation by 8-OH-DPAT (Eriksson et al., 2012a). Eriksson et al. (2012a) proposed that dorsohippocampal infusion of 8-OH-DPAT impaired passive avoidance retention through hippocampal 5-HT$_{1A}$ receptor activation, while 5-HT$_7$ receptor appears to facilitate memory processes in a broader cortico-limbic network and not the hippocampus alone. Importantly, Huang et al. (2012) reported that 5-HT$_{1A}$ and 5-HT$_7$ receptors contribute to lurasidone-induced dopamine efflux, concluding that lurasidone (atypical antipsychotic) on the PFC and hippocampus, DA efflux are dependent, at least partially, on its 5-HT$_{1A}$ agonist and 5-HT$_7$ antagonist properties and may contribute to its efficacy to reverse the effects of subchronic phencyclidine treatment and improve schizophrenia. Moreover, Freret et al. (2013) reported that with a 2-h delay, SB-269970 (3.0 and 10.0 mg/kg, administered subcutaneously) impaired the discrimination of the novel object. With a 4-h delay, while control mice were not able to discriminate the novel object, mice treated with 5-carboxamidotryptamine (5-CT, a $5\text{-HT}_{1A/1B/1D/7}$ receptors agonist; 1.0 mg/kg) showed a significant discrimination; this promnesic effect with a long delay is effectively mediated by 5-HT_7 receptor activation since it was blocked by SB-269970 (10.0 mg/kg), but not by WAY100135 (10.0 mg/kg) or by GR-127935 (10.0 mg/kg) (Freret et al., 2013).

Of course, the analysis of 5-HT_7 receptor role in learning and memory had further enriched by evidence that 8-OH-DPAT dual affinity (see above) and functional effect involving 5-HT_{1A} and 5-HT_7 receptors in different protocols of training/testing (Perez-Garcia and Meneses, 2005b) and neuroprotective effect (Malá et al., 2013). An interesting further complication is the evidence, as above mentioned, about the heterodimerization of serotonin receptors 5-HT_{1A} and 5-HT_7 differentially regulating receptor signaling and trafficking (Renner et al., 2012).

Importantly, according to Matthys et al. (2011), pharmacological and genetic tools targeting the 5-HT_7 receptor in preclinical animal models have implicated it in diverse (patho)physiological processes of the CNS and data obtained with 5-HT_7 receptor KO mice, selective antagonists, and, to a lesser extent, agonists, however, are quite contradictory. Matthys et al. (2011) and Gellynck et al. (2013) discuss in detail the role of the 5-HT_7 receptor in the CNS and propose some hypothetical models, which could explain the observed inconsistencies. These models are based on two novel concepts within the field of GPCRs, namely biphasic signaling and G protein-independent signaling, which both have been shown to be mediated by GPCR dimerization (Matthys et al., 2011). Notably, these same authors suggest that the 5-HT_7 receptor could reside in different dimeric contexts and initiate different signaling pathways, depending on the neuronal circuitry and/or brain region; concluding GPCR dimerization and G protein-independent signaling as two promising future directions in 5-HT_7 receptor research, which ultimately might lead to the development of more efficient dimer- and/or pathway-specific therapeutics (Matthys et al., 2011).

It should be noted that, e.g., a drug (Egis-11150) displaying affinity and intrinsic activity for diverse neurotransmission systems, including inverse agonism for 5-HT_7 receptor had a robust procognitive profile in preclinical models (Gacsályi et al., 2013) and according to them a potential treatment for controlling the psychosis as well as the cognitive dysfunction in schizophrenia. Two other notable 5-HT$_7$ receptor inverse agonists present interesting features. For instance, clozapine displays affinity for several receptors and neurotransmission systems (Meltzer et al., 2012), including dopamine D$_3$/D$_4$, 5-HT$_{1A}$/5-HT$_{2A}$/5-HT$_6$/5-HT$_7$ receptors, cholinergic, histaminergic, and adrenergic systems. It should be noted that clozapine has several side effects (Lyseng-Williamson, 2013); notwithstanding, important in the present context, the improvement in positive and negative symptoms, general psychopathology, cognition, suicidality/mood, and fewer extrapyramidal side effects (EPS) contribute to clozapine having broader advantages that most probably represent the integrated sum of these effects (Meltzer, 2012). According to Meltzer (2012), these include improvement in work and social function, quality of life, lower relapse rate, and rehospitalization; much of this depends upon improvement in cognitive function. Likewise, clozapine has been reliably shown to improve some domains of cognition in schizophrenia, including verbal fluency, declarative memory, attention, and speeded mental functions (Meltzer, 2012).

Notably, Smith et al. (2006) reported that risperidone displays a novel mechanism of antagonism of the h5-HT$_7$ receptor, interacting in an irreversible or pseudo-irreversible manner with the h5-HT$_7$ receptor, thus producing the inactivation. Internalization of the h5-HT$_7$ receptor was not detected by monitoring green fluorescent protein-labeled fluorescent forms of the h5-HT$_7$ receptor exposed to risperidone and 10 other antagonists tested for h5-HT$_7$ inactivating properties, and only 9-OH-risperidone and methiothepin (which displays affinity for diverse 5-HT receptors; Hoyer et al., 1994; McLoughlin and Strange, 2000) were found to demonstrate the same anomalous properties as risperidone (Smith et al., 2006). According to Smith et al. (2006) their results indicate that the h5-HT$_7$ receptor may possess unique structural features that allow certain drugs to induce a conformation resulting in an irreversible interaction in the intact membrane environment, which may indicate that the h5-HT$_7$ receptor is part of a subfamily of GPCRs possessing this property or that many GPCRs have the potential to be irreversibly blocked, but only selected drugs can induce this effect. At the very least, the possibility that highly prescribed drugs, such as risperidone, are irreversibly antagonizing GPCR function *in vivo* is noteworthy (Smith et al., 2006). It should be noted that according to McIntosh et al. (2013), risperidone only partially reverses the schizophrenic symptomology; since it reversed some, but not all, of the learning and memory deficits induced by postweaning isolation model, the isolation rearing model may be useful to predict antipsychotic activity of novel therapeutic agents. According to Goghari et al. (2013), in humans no changes were found in spatial WM ability and short-term atypical treatment with risperidone or quetiapine can increase prefrontal cortical thickness in psychosis. Moreover, Smith et al. (2011) noted risperidone-induced inactivation and clozapine-induced reactivation of rat cortical astrocyte $5-HT_7$ receptors, thus providing evidence for *in situ* GPCR homodimer protomer cross talk.

Tarazi and Riva (2013) provided other notable example. Lurasidone is a novel antipsychotic drug approved for the treatment of schizophrenia in adults, which demonstrated high affinity for $5-HT_{1A}$, $5-HT_{2A}$, $5-HT_7$, dopamine D_2, and adrenergic α_{2C} receptors followed by α_1 and α_{2A} receptors and the drug was active in animal models predictive of antipsychotic and antidepressant activities; in addition, it demonstrated procognitive effects, as it was effective in several animal models that assessed memory, cognition, and executive functions in rats and in primates. Importantly, according to Tarazi and Riva (2013) at a cellular level, lurasidone promotes neuronal plasticity, can modulate epigenetic mechanisms controlling gene transcription, and increases the expression of the neurotrophic factor BDNF (brain-derived neurotrophic factor) in cortical and limbic brain regions. These authors conclude that the mechanisms of action of lurasidone might contribute to its unique psychopharmacological properties in the improved treatment of schizophrenia and perhaps other psychiatric disorders (Tarazi and Riva, 2013).

Relevant to the present context, the review by Shimizu et al. (2013) highlights that although a series of second generation antipsychotics (SGAs) (e.g., risperidone, olanzapine, and quetiapine) have been developed in the past two decades, clinical reports do not necessarily show advantages over first generation antipsychotics (FGAs) in the treatment of schizophrenia, especially in their efficacy against cognitive impairment and ability to cause EPS. Recently, several lines of studies have revealed therapeutic roles of 5-HT receptors in modulating cognitive impairments and extrapyramidal motor disorders (Shimizu et al., 2013). Inhibition of $5-HT_{1A}$, $5-HT_3$, and $5-HT_6$ receptors or activation of $5-HT_4$ receptors alleviates cognitive impairments (e.g., deficits in learning and memory); in addition, stimulation of $5-HT_{1A}$ receptors or inhibition of $5-HT_3$ and $5-HT_6$ receptors as well as $5-HT_{2A/2C}$ receptors can ameliorate extrapyramidal motor disorders (Shimizu et al., 2013). Thus, controlling the activity of 5-HT$_{1A}$, 5-HT$_3$, or 5-HT$_6$ (even 5-HT$_7$) receptors seems to provide benefits by both alleviating cognitive impairments and reducing antipsychotic-induced EPS (Shimizu et al., 2013). It should be noted that while methiothepin and SB-269970 displayed similar negative intrinsic activity to SB-691673 at the rat 5-HT$_{7A}$ receptor (Romero et al., 2006), the compounds SB-258719, mesulergine, and metergoline displayed some lower negative intrinsic activity. With the exception of SB-258719 and mesulergine, which remained a partial inverse agonist at the human 5-HT$_{7A}$ receptor, the other compounds behaved with a similar Emax value to the full inverse agonist SB-691673. In conclusion, none of the 5-HT receptor antagonists investigated displayed silent properties at the rat or human 5-HT$_{7A}$ receptor, when these are expressed in a system allowing detection of constitutive activity (Romero et al., 2006). Romero et al (2004) conclude that apparently to be partial to full inverse agonists, further illustrating that an antagonist is preferentially an inverse agonist when investigated under constitutively active receptor conditions (Romero et al., 2006).

12.1 SERT

Importantly, the analysis of the interaction among brain areas and neurotransmitter systems (Briand et al., 2007) shows that the expression of diverse transporters (e.g., GABA, glutamate, serotonin) is modulated by memory formation, amnesia, or forgetting (Tellez et al., 2012a,b). It should be noted that transporters control neurotransmission systems by removing the neurotransmitter from the extracellular space. Importantly, regarding SERT, Timotijević et al. (2012) had highlighted that serotonergic antidepressants by changing the concentration of serotonin alter primarily affective manifestations; they also have significant influence on all the spectrum of serotonergic disorders not only emotional but also the cognitive level, which is also a confirmation that the therapeutic effects do not depend only on the simple change of serotonin concentration but also on the level where these changes occur in dynamic comparison of key transmitters. Hence, serotonergic mechanisms seem to be involved in memory functions and dysfunctions.

Moreover, the SERT has been associated with diverse functions and diseases, though seldom to memory. Therefore, an attempt was made to summarize and discuss the available publications implicating the involvement of the SERT in memory, amnesia, and antiamnesic effects (Meneses, 2013; Meneses et al., 2011b). Evidence indicates that AD and drugs of abuse like D-methamphetamine (METH) (Tellez et al., 2012a,b) and (+/−)3,4-methylenedioxymethamphetamine (MDMA, "ecstasy") have been associated to decrements in the SERT expression and memory deficits (Meneses et al., 2011b; Parrott, 2013a,b). Also, memory formation and amnesia affected the SERT expression; hence, the SERT expression seems to be a reliable neural marker related to memory mechanisms, its alterations and potential treatment. Notably, individuals with low 5-HTT expression performed significantly better on a test of memory compared to individuals with medium 5-HTT expression, suggesting that possession of low-expressing genetic variants of 5-HTT is modestly associated with enhanced cognitive performance among healthy older adults (Salminen et al., 2013). Tellez et al. (2010) reported that trained animals decreased cortical SERT binding relative to untrained ones; in untrained and trained treated animals with the amnesic dose of METH SERT binding in several areas including hippocampus and cortex decreased, more remarkably in the trained animals. In contrast, fluoxetine improved memory, increased SERT binding, prevented the METH amnesic effect, and reestablished the SERT binding. In general, memory and amnesia seemed to make SERT more vulnerable to drug effects (Tellez et al., 2010). Even SERT seems to be associated to forgetting (Tellez et al., 2012b).

The pharmacological, neural, and molecular mechanisms associated with these changes are of great importance for investigation (Meneses et al., 2011b). For instance, high tryptophan diet reduces hippocampal CA1 intraneuronal β-amyloid in the triple transgenic mouse model of AD (Noristani et al., 2012), probably, involving SERT. Importantly, verbal memory deficits are correlated with prefrontal hypometabolism in recreational MDMA users (Bosch et al., 2013); however, Chou et al. (2012) investigated the impaired cognition in bipolar disorder, finding that the overall deficits in cognition were not significantly correlated with the SERT availability or the brain-derived neurotrophic factor. Importantly, brain-specific tryptophan hydroxylase and 5-HTTLPR (SERT gene) are associated with frontal lobe symptoms in AD (Engelborghs et al., 2013). Hence, likely, SERT and memory investigation will have a long history. For instance, Chow et al. (2007) had highlighted the potential cognitive enhancing and disease modification effects of SSRIs for AD; certainly, regarding SSRIs and young population caution is important (Karanges et al., 2013).

Burghardt and Bauer (2013) argue that by altering activity in specific brain areas, acute SSRI administration enhances both acquisition and expression of cued fear conditioning, by reducing activity/plasticity within the hippocampus. Burghardt and Bauer (2013) proposed that both acute and chronic SSRI treatment impairs the expression of context fear conditioning and finally, the impairments in fear learning and memory found with chronic SSRI treatment can be attributed to changes in glutamatergic neurotransmission in the amygdala and hippocampus.

Harmer and Cowen (2013) highlight that both behavioral and neuro-imaging studies show that SSRI administration produces positive biases in attention, appraisal, and memory from the earliest stages of treatment, well before the time that clinical improvement in mood becomes apparent, suggesting that the delay in the clinical effect of SSRIs can be explained by the time needed for this positive bias in implicit emotional processing to become apparent at a subjective, conscious level. According to Harmer and Cowen (2013), this process likely involves the relearning of emotional associations in a new, more positive emotional environment, suggesting intriguing links between the effect of SSRIs to promote synaptic plasticity and neurogenesis, and their ability to remediate negative emotional biases in depressed patients. Interestingly, like the naïve 5-HTT($-/-$) rats, the cocaine exposed 5-HTT($-/-$) rats displayed improved cognitive flexibility (Nonkes et al., 2013). These same authors conclude that improved reversal learning in 5-HTT($-/-$) rats reflects a pre-existing trait that is preserved during cocaine-withdrawal; as 5-HTT ($-/-$) rodents model the low activity s-allele of the human serotonin transporter-linked polymorphic region, these findings may have heuristic value in the treatment of s-allele cocaine addicts (Nonkes et al., 2013). Even the role of serotonin in cognitive function is being supported by recent evidence with implications for understanding depression (Cowen and Sherwood, 2013) or signaling (Fournet et al., 2012). For instance, a chicken essence acting as a dual inhibitor of the SERT and acetylcholinesterase significantly shortened escape latency in the water maze test in depressed mice previously subjected to a repeated open-space swimming task, which induces a depression-like state (Tsuruoka et al., 2012).

According to Kuhn et al. (2013), a vast number of imaging studies have demonstrated the impact of 5-HT and BDNF on emotion and memory-related networks in the context of MDD. Underlying molecular mechanisms that affect the functionality of these networks have been examined in detail in animals and corroborate imaging findings. The crucial role of 5-HT and BDNF signaling in the context of MDD is reflected in the etiologic models of MDD such as the monoamine or neuroplasticity hypothesis as well as in pharmacological models of antidepressant response. While antidepressant drug treatment has been primarily linked to the modulation of emotion-related networks, cognitive behavioral therapy has been implicated in a top-down control of limbic structures. According to Kuhn et al. (2013), a simple lack of monoamines or BDNF has been proposed as causal factor of MDD etiology; however, recent findings suggest a much more complex neurobiology emphasizing epistatic and epigenetic mechanisms responsible for structural and functional changes observed in emotion and memory-related brain regions of healthy subjects and MDD patients. Kuhn et al. (2013) review focuses on neuroimaging studies in the context of MDD, and they provide a comprehensive overview of these networks as well as on the specific role of 5-HT and BDNF in their development and function.

Certainly, BDNF and neural plasticity have been linked to 5-HT (Tarazi and Riva (2013) or other neurotransmission systems (e.g., glutamate; Lynch et al., 2008). It should be noted that Karanges et al. (2013) reported that chronic administration of the SSRI paroxetine in adult and adolescent rats produces age-specific changes in the hippocampal proteome (e.g., PKA, probably affecting CREB, BDNF). Although similar changes were observed in many proteins in both age groups, there were notable differences in the expression profiles of proteins implicated in apoptosis, oxidative stress, cytoskeletal structure, intracellular signaling, and serotonergic and catecholaminergic neurotransmission (Karanges et al., 2013). According to Karanges et al. (2013), their findings, while suggestive rather than conclusive, demonstrate that the developing brain responds to paroxetine in a manner distinct from the adult brain and provides some clues to the mechanisms underlying the adolescent (in contrast with adult) response to antidepressant drugs. Nonkes et al. (2014) studied the hypothesis that dopamine, serotonin mediate individual differences in sensitivity to CS as measured in the sign- versus goal-tracking task.

CONCLUSIONS

Certainly why and how $5\text{-HT}_{1A/1B}$, $5\text{-HT}_{2A/2B/2C}$, 5-HT_4, and 5-HT_6, 5-HT_7 receptor agonists and/or antagonists or SSRIs may facilitate memory or reverse amnesia in some memory tasks (Huerta-Rivas et al., 2010; King et al., 2008; Upton et al., 2008) left open very interesting possibilities. Doubtless, further experiments are necessary to clarify the role of 5-HT systems during memory formation, e.g., testing diverse compounds (agonists and antagonists, inverse agonists, SSRIs), neural markers, and behavioral tasks, which might offer important insights about memory formation and its alterations. Probably, the development of effective treatments for memory alterations has been limited by the absence of reliable markers to indicate and/or predict efficacy. This is important in the context as in as much as human studies suggest a potential utility of some (Geldenhuys and Van der Schyf, 2009; Reid et al., 2010; Upton et al., 2008). For instance, this premise has been translated into the clinical efficacy of some 5-HT_6 receptor antagonists in mild-to-moderate AD patients (Liu and Robichaud, 2009; Rossé and Schaffhauser, 2010), and even individuals with mild cognitive impairment offer a great opportunity. Importantly, the number of pharmacologically and structurally distinct 5-HT receptors (Barnes, 2011) is 13 for humans, but even this number of different receptors is not the whole story. This monoamine is an ancient biochemical derived from the amino acid, tryptophan, and it is widely used throughout the animal and plant kingdoms and has evolved mechanisms that impact cell biology—either directly or indirectly—in increasingly recognized ways in addition to the "classical" activation of cell membrane receptors (Barnes, 2011). In addition, Millan (2011) highlights microRNA in the regulation and expression of serotonergic transmission in the brain and other tissues (Millan, 2011).

The role of 5-HT systems on memory and dysfunctional memory in diverse psychiatric diseases has become a major area of scientific interest, being 5-HT_{1A}, 5-HT_{1B}, 5-HT_{2A-2C}, 5-HT_4, 5-HT_6 receptors or SERT the more recent focus; certainly, 5-HT_3 receptor cannot be excluded. Clearly available data show the interaction among 5-HT systems, memory, amnesia, and prevention of the amnesia. Further interest is provided by the evidence that both 5-HT receptor agonists and antagonists may have promnesic and/or antiamnesic effects in conditions covering memory formation, age-related cognitive impairments, memory deficits in diseases (e.g., drug addition, PTSD, schizophrenia, Parkinson, AD). Why some 5-HT receptor agonists and antagonists may have promnesic and/or antiamnesic effects is unclear, but this situation is not new regarding 5-HT$_{1A}$ receptor. Inverse agonists and memory have been mentioned (Meneses, 1999; Wallace et al., 2011), particularly regarding GABAergic (Sarter and Stephens, 1988) and histaminergic (Alleva et al., 2013) systems, benzodiazepines (Izquierdo et al., 1990), 5-HT$_{1A}$ receptor (McLoughlin and Strange, 2000; Meneses and Perez-Garcia, 2007), 5-HT$_{1B}$ receptor (Meneses, 2001), 5-HT$_{2A}$ receptor (Aloyo et al., 2009), 5-HT$_4$ receptor (Bockaert et al., 1998, 2008), 5-HT$_6$ receptor (Meneses et al., 2011a,b), ion channel modulation of neurotransmission (Eid and Rose, 1999), and cannabinoid receptors (Reggio, 2003). Moreover, one way to address important questions is by continuing testing 5-HT receptor agonists, antagonists, and inverse agonists as well as SSRIs in memory tasks and to analyze changes in brain areas and neurobiological markers (e.g., 5-HT$_6$ receptor protein or mRNA expression, signaling cascades), in animals under memory formation or amnesia, forgetting plus drugs administration. The identification of reliable neural markers is fundamental for the understanding of memory mechanisms, its alterations, and potential treatment and possible cure? The memory and molecular changes might represent new insights and promise steps, mainly in the light of pharmacological magnetic resonance imaging (phMRI; Martin and Sibson, 2008; Meyer, 2012; Zimmer and Le Bars, 2013), which offers potential novel insights into the functioning of neurotransmitter systems and drug action in the CNS and memory formation, amnesia, forgetting, or behavioral/psychiatric alterations. Such studies might be useful in providing markers of the neuropharmacological modulation of neuronal activity across the whole brain with spatial and temporal specificity (SERT seems to be a good candidate), which might provide useful markers for diagnoses. Also further investigation using different memory tasks, times, and amnesia models might provide important clues. Improving devices for measuring behavioral memory is crucial. The potential usefulness of 5-HT markers for treatment, monitoring, and in the process of developing and evaluating novel drugs is worthy of investigation; as it is offering new alternative venues for memory investigation and dysfunctional memory in diverse psychiatric disorders (see e.g., Millan et al., 2012; Meneses, 2013; 2014). Available evidence offers clues about possible but the exact mechanisms remain unclear.

REFERENCES

Alleva, L., Tirelli, E., Brabant, C., 2013. Therapeutic potential of histaminergic compounds in the treatment of addiction and drug-related cognitive disorders. Behav. Brain Res. 237, 357–368.

Aloyo, V.J., Berg, K.A., Spampinato, U., Clarke, W.P., Harvey, J.A., 2009. Current status of inverse agonism at serotonin2A (5-HT_{2A}) and 5-HT_{2C} receptors. Pharmacol. Ther. 121 (2), 160–173.

Andrade, R., Haj-Dahmane, S., 2013. Serotonin neuron diversity in the dorsal raphe. ACS Chem. Neurosci. 4 (1), 22–25. Available from: http://dx.doi.org/10.1021/cn300224n. PMID: 23336040.

Aubert, Y., Allers, K.A., Sommer, B., de Kloet, E.R., Abbott, D.H., Datson, N.A., 2013. Brain region-specific transcriptomic markers of serotonin-1A receptor agonist action mediating sexual rejection and aggression in female marmoset monkeys. J. Sex Med. 10 (6), 1461–1475.

Ballaz, S.J., Ail, H., Watson, S.J., 2007. Analysis of 5-HT_6 and 5-HT_7 receptor gene expression in rats showing differences in novelty-seeking behavior. Neuroscience 147 (2), 428–438.

Barnes, N.M., 2011. 5-HT: the promiscuous and happy hormone!. Curr. Opin. Pharmacol. 11 (1), 1–2.

Barnes, N.M., Hales, T.G., Lummis, S.C., Peters, J.A., 2009. The 5-HT_3 receptor—the relationship between structure and function. Neuropharmacology 56 (1), 273–284.

Batsikadze, G., Paulus, W., Kuo, M.F., Nitsche, M.A., 2013. Effect of serotonin on paired associative stimulation-induced plasticity in the human motor cortex. Neuropsychopharmacology. Available from: http://dx.doi.org/10.1038/npp.2013.127. [Epub ahead of print] PMID: 23680943.

Belcher, A.M., O'Dell, S.J., Marshall, J.F., 2005. Impaired object recognition memory following methamphetamine, but not p-chloroamphetamine- or d-amphetamine-induced neurotoxicity. Neuropsychopharmacology 30 (11), 2026–2034.

Blasi, G., De Virgilio, C., Papazacharias, A., Taurisano, P., Gelao, B., Fazio, L., et al., 2013. Converging evidence for the association of functional genetic variation in the serotonin receptor 2a gene with prefrontal function and olanzapine treatment. JAMA Psychiatry. Available from: http://dx.doi.org/10.1001/jamapsychiatry.2013.1378 [Epub ahead of print] PMID: 23842608.

Bockaert, J., Claeysen, S., Sebben, M., Dumuis, A., 1998. 5-HT_4 receptors: gene, transduction and effects on olfactory memory. Ann. N. Y. Acad. Sci. 861, 1–15.

Bockaert, J., Claeysen, S., Bécamel, C., Dumuis, A., Marin, P., 2006. Neuronal 5-HT metabotropic receptors: fine-tuning of their structure, signaling, and roles in synaptic modulation. Cell Tissue Res. 326, 553–572.

Bockaert, J., Claeysen, S., Compan, V., Dumuis, A., 2008. 5-HT_4 receptors: history, molecular pharmacology and brain functions. Neuropharmacology 55, 922–931.

Bockaert, J., Perroy, J., Bécamel, C., Marin, P., Fagni, L., 2010. GPCR interacting proteins (GIPs) in the nervous system: roles in physiology and pathologies. Annu. Rev. Pharmacol. Toxicol. 50, 89–109.

Bockaert, J., Claeysen, S., Compan, V., Dumuis, A., 2011. 5-HT_4 receptors, a place in the sun: act two. Curr. Opin. Pharmacol. 11 (1), 87–93.

Boess, F.G., de Vry, J., Erb, C., Flessner, T., Hendrix, M., Luithle, J., et al., 2013. Pharmacological and behavioral profile of N-[(3R)-1-azabicyclo[2.2.2]oct-3-yl]-6-chinolincarboxamide (EVP-5141), a novel $\alpha7$ nicotinic acetylcholine receptor agonist/serotonin 5-HT_3 receptor antagonist. Psychopharmacology (Berlin) 227 (1), 1–17.

Bombardi, C., Di Giovanni, G., 2013. Functional anatomy of 5-HT$_{2A}$ receptors in the amygdala and hippocampal complex: relevance to memory functions. Exp. Brain Res. [Epub ahead of print] PMID: 23591691.

Borg, J., 2008. Molecular imaging of the 5-HT$_{1A}$ receptor in relation to human cognition. Behav. Brain Res. 195 (1), 103–111.

Borroni, B., Costanzi, C., Padovani, A., 2010. Genetic susceptibility to behavioural and psychological symptoms in Alzheimer disease. Curr. Alzheimer Res. 7 (2), 158–164.

Bosch, O.G., Wagner, M., Jessen, F., Kühn, K.U., Joe, A., Seifritz, E., et al., 2013. PET of recreational MDMA users. PLoS One 8 (4), e61234.

Boulougouris, V., Robbins, T.W., 2010. Enhancement of spatial reversal learning by 5-HT$_{2C}$ receptor antagonism is neuroanatomically specific. J. Neurosci. 30 (3), 930–938.

Briand, L.A., Gritton, H., Howe, W.M., Young, D.A., Sarter, M., 2007. Modulators in concert for cognition: modulator interactions in the prefrontal cortex. Prog. Neurobiol. 83 (2), 69–91.

Buhot, M.C., Wolff, M., Benhassine, N., Costet, P., Hen, R., Segu, L., 2003a. Spatial learning in the 5-HT$_{1B}$ receptor knockout mouse: selective facilitation/impairment depending on the cognitive demand. Learn. Mem. 10 (6), 466–477.

Buhot, M.C., Wolff, M., Savova, M., Malleret, G., Hen, R., Segu, L., 2003b. Protective effect of 5-HT$_{1B}$ receptor gene deletion on the age-related decline in spatial learning abilities in mice. Behav. Brain Res. 142 (1-2), 135–142.

Buhot, M.C., Wolff, M., Segu, L., 2003c. Serotonin. In: Riedel, G., Platt, B. (Eds.), Memories are Made of These: From Messengers to Molecules. Eurekah.com and Kluwer Academic/Plenum Publishers, Georgetown, TX, pp. 1–19.

Burghardt, N.S., Bauer, E.P., 2013. Acute and chronic effects of selective serotonin reuptake inhibitor treatment on fear conditioning: implications for underlying fear circuits. Neuroscience 247, 253–272.

Bussey, T.J., Holmes, A., Lyon, L., Mar, A.C., McAllister, K.A., Nithianantharajah, J., et al., 2012. New translational assays for preclinical modelling of cognition in schizophrenia: the touchscreen testing method for mice and rats. Neuropharmacology 62 (3), 1191–1203.

Cahir, M., Ardis, T., Reynolds, G.P., Cooper, S.J., 2007. Acute and chronic tryptophan depletion differentially regulate central 5-HT$_{1A}$ and 5-HT$_{2A}$ receptor binding in the rat. Psychopharmacology (Berlin) 190 (4), 497–506.

Cahir, M., Ardis, T.C., Elliott, J.J., Kelly, C.B., Reynolds, G.P., Cooper, S.J., 2008. Acute tryptophan depletion does not alter central or plasma brain-derived neurotrophic factor in the rat. Eur. Neuropsychopharmacol. 18 (5), 317–322.

Cai, X., Kallarackal, A.J., Kvarta, M.D., Goluskin, S., Gaylor, K., Bailey, A.M., et al., 2013. Local potentiation of excitatory synapses by serotonin and its alteration in rodent models of depression. Nat. Neurosci. 16 (4), 464–472.

Calcagno, E., Carli, M., Invernizzi, R.W., 2006. The 5-HT$_{1A}$ receptor agonist 8-OH-DPAT prevents prefrontocortical glutamate and serotonin release in response to blockade of cortical NMDA receptors. J. Neurochem. 96 (3), 853–860.

Callaghan, C.K., Hok, V., Della-Chiesa, A., Virley, D.J., Upton, N., O'Mara, S.M., 2012. Age-related declines in delayed non-match-to-sample performance (DNMS) are reversed by the novel 5HT$_6$ receptor antagonist SB742457. Neuropharmacology 63 (5), 890–897.

Campan, V., Zhou, M., Grailhe, R., Gazzara, R.A., Martin, R., Gingrich, J., et al., 2004. Attenuated responses to stress and novelty and hypersensitivity to seizures in 5-HT$_4$ receptor knock-out mice. J. Neurosci. 24, 412–419.

Carli, M., Samanin, R., 1992. 8-Hydroxy-2-(di-n-propylamino) tetralin impairs spatial learning in a water maze: role of postsynaptic 5-HT$_{1A}$. Br. J. Pharmacol. 105, 720–726.

Carli, M., Samanin, R., 2000. The 5-HT$_{1A}$ receptor agonist 8-OH-DPAT reduces rats' accuracy of attentional performance and enhances impulsive responding in a five-choice serial reaction time task: role of presynaptic 5-HT$_{1A}$ receptors. Psychopharmacology (Berlin) 149, 259–268.

Carli, M., Luschi, R., Garofalo, P., Samanin, R., 1995. 8-OH-DPAT impairs spatial but not visual learning in a water maze by stimulating 5-HT$_{1A}$ receptors in the hippocampus. Behav. Brain Res. 67, 67–74.

Carli, M., Bonalumi, P., Samanin, R., 1997. WAY 100635, a 5-HT$_{1A}$ receptor antagonist, prevents the impairment of spatial learning caused by intrahippocampal administration of scopolamine or 7-chloro-kynurenic acid. Brain Res. 774, 167–174.

Carli, M., Bonalumi, P., Samanin, R., 1998. Stimulation of 5-HT$_{1A}$ receptors in the dorsal raphe reverses the impairment caused by intrahippocampal scopolamine in rats. Eur. J. Neurosci. 10, 221–230.

Carli, M., Silva, S., Balducci, C., Samanin, R., 1999a. WAY 100635, a 5-HT1A receptor antagonist, prevents the impairment of spatial learning caused by blockade of hippocampal NMDA receptors. Neuropharmacology 38, 1165–1173.

Carli, M., Balducci, C., Millan, M.J., Bonalumi, P., Samanin, R., 1999b. S 15535, a benzodioxopiperazine acting as presynaptic agonist and postsynaptic 5-HT$_{1A}$ receptor antagonist, prevents the impairment of spatial learning caused by intrahippocampal scopolamine. Br. J. Pharmacol. 128, 1207–1214.

Carli, M., Balducci, C., Samanin, R., 2001. Stimulation of 5-HT$_{1A}$ receptors in the dorsal raphe ameliorates the impairment of spatial learning caused by intrahippocampal 7-chloro-kynurenic acid in naive and pretrained rats. Psychopharmacology (Berlin) 158 (1), 39–47.

Carli, M., Baviera, M., Invernizzi, R.W., Balducci, C., 2006. Dissociable contribution of 5-HT$_{1A}$ and 5-HT$_{2A}$ receptors in the medial prefrontal cortex to different aspects of executive control such as impulsivity and compulsive perseveration in rats. Neuropsychopharmacology 31, 757–767.

Cassel, J.C., 2010. Experimental studies on the role(s) of serotonin in learning and memory functions. In: Muller, C.P., Jacobs, B.L. (Eds.), Handbook of the Behavioral Neurobiology of Serotonin, vol. 21. Academic Press, Amsterdam.

Castillo, C., Ibarra, M., Márquez, J.A., Villalobos-Molina, R., Hong, E., 1993. Pharmacological evidence for interactions between 5-HT$_{1A}$ receptor agonists and subtypes of alpha 1-adrenoceptors on rabbit aorta. Eur. J. Pharmacol. 241 (2–3), 141–148.

Chegini, H.R., Nasehi, M., Zarrindast, M.R., 2013. Differential role of the basolateral amygdala 5-HT3 and 5-HT4 serotonin receptors upon ACPA-induced anxiolytic-like behaviors and emotional memory deficit in mice. Behav. Brain Res.pii: S0166-4328(13)00750-X. doi:10.1016/j.bbr.2013.12.007. [Epub ahead of print] PMID: 24333573 [PubMed—as supplied by publisher].

Chou, Y.H., Wang, S.J., Lirng, J.F., Lin, C.L., Yang, K.C., Chen, C.K., et al., 2012. Impaired cognition in bipolar I disorder: the roles of the serotonin transporter and brain-derived neurotrophic factor. J. Affect. Disord. 143 (1–3), 131–137.

Chow, T.W., Pllock, B.G., Milgram, N.W., 2007. Potential cognitive enhancing and disease modification effects of SSRIs for Alzheimer's disease. Neuropsychiatr. Dis. Treat. 3, 627–636.

Cifariello, A., Pompili, A., Gasbarri, A., 2008. 5-HT$_7$ receptors in the modulation of cognitive processes. Behav. Brain Res. 195 (1), 171–179.

Cochet, M., Donneger, R., Cassier, E., Gaven, F., Lichtenthaler, S.F., Marin, P., et al., 2013. 5-HT$_4$ receptors constitutively promote the non-amyloidogenic pathway of APP cleavage and interact with ADAM10. ACS Chem. Neurosci. 4, 130–140.

Cook, R.G., Geller, A.I., Zhang, G.R., Gowda, R., 2004. Touchscreen-enhanced visual learning in rats. Behav. Res. Meth. Instrum. Comput. 36 (1), 101–106.

Costall, B., 1993. The breadth of action of the 5-HT$_3$ receptor antagonists. Int. Clin. Psychopharmacol. 8 (Suppl. 2), 3–9.

Cowen, P., Sherwood, A.C., 2013. The role of serotonin in cognitive function: evidence from recent studies and implications for understanding depression. J. Psychopharmacol. [Epub ahead of print] PMID: 23535352.

Curtin, P.C.P., Medan, V., Neumeister, H., Bronson, D.R., Preuss, T., 2013. The 5-HT$_{5A}$ receptor regulates excitability in the auditory startle circuit: functional implications for sensorimotor gating. J. Neurosci. 33 (24), 10011–10020.

Da Silva Costa-Aze, V., Quiedeville, A., Boulouard, M., Dauphin, F., 2012. 5-HT$_6$ receptor blockade differentially affects scopolamine-induced deficits of working memory, recognition memory and aversive learning in mice. Psychopharmacology (Berlin) 222 (1), 99–115.

Dayan, P., Huys, Q.J., 2009. Serotonin in affective control. Annu. Rev. Neurosci. 32, 95–126.

van Donkelaar, E.L., Blokland, A., Ferrington, L., Kelly, P.A., Steinbusch, H.W., Prickaerts, J., 2011. Mechanism of acute tryptophan depletion: is it only serotonin? Mol. Psychiatry 16 (7), 695–713.

Dougherty, J.P., Oristaglio, J., 2013. Chronic treatment with the serotonin 2A/2C receptor antagonist SR 46349B enhances the retention and efficiency of rule-guided behavior in mice. Neurobiol. Learn. Mem.. Available online April 12, 2013; doi:/10.1016/j.nlm. 2013.04.002.

Drago, A., Alboni, S., Brunello, N., De Ronchi, D., Serretti, A., 2010. HTR1B as a risk profile maker in psychiatric disorders: a review through motivation and memory. Eur. J. Clin. Pharmacol. 66 (1), 5–27 [Epub October 7, 2009]. Review. Erratum in: Eur. J. Clin. Pharmacol., 66 (1), 105.

Eid Jr, C.N., Rose, G.M., 1999. Cognition enhancement strategies by ion channel modulation of neurotransmission. Curr. Pharm. Des. 5 (5), 345–361.

Elvander-Tottie, E., Eriksson, T.M., Sandin, J., Ogren, S.O., 2009. 5-HT$_{1A}$ and NMDA receptors interact in the rat medial septum and modulate hippocampal-dependent spatial learning. Hippocampus 12 (19), 1187–1198.

Engelborghs, S., Sleegers, K., Van der Mussele, S., Le Bastard, N., Brouwers, N., Van Broeckhoven, C., et al., 2013. Brain-specific tryptophan hydroxylase, TPH2, and 5-HTTLPR are associated with frontal lobe symptoms in Alzheimer's disease. J. Alzheimers Dis. [Epub ahead of print] PMID: 23334703.

Eppinger, B., Hämmerer, D., Li, S.C., 2012. Neuromodulation of reward-based learning and decision making in human aging. Ann. N. Y. Acad. Sci. 1235, 1–17.

Eriksson, T.M., Madjid, N., Elvander-Tottie, E., Stiedl, O., Svenningsson, P., Ogren, S.O., 2008. Blockade of 5-HT$_{1B}$ receptors facilitates contextual aversive learning in mice by disinhibition of cholinergic and glutamatergic neurotransmission. Neuropharmacology 54 (7), 1041–1050.

Eriksson, T.M., Holst, S., Stan, T.L., Hager, T., Sjögren, B., Ogren, S.O., et al., 2012a. 5-HT$_{1A}$ and 5-HT$_7$ receptor crosstalk in the regulation of emotional memory: implications for effects of selective serotonin reuptake inhibitors. Neuropharmacology 63 (6), 1150–1160.

Eriksson, T.M., Alvarsson, A., Stan, T.L., Zhang, X., Hascup, K.N., Hascup, E.R., et al., 2012b. Bidirectional regulation of emotional memory by 5-HT$_{1B}$ receptors involves hippocampal p11. Mol. Psychiatry. Available from: http://dx.doi.org/10.1038/mp.2012.130. [Epub ahead of print].

Eriksson, T.M., Delagrange, P., Spedding, M., Popoli, M., Mathé, A.A., Ögren, S.O., et al., 2012c. Emotional memory impairments in a genetic rat model of depression: involvement of 5-HT/MEK/Arc signaling in restoration. Mol. Psychiatry 17 (2), 173–184.

Ersche, K.D., Roiser, J.P., Lucas, M., Domenici, E., Robbins, T.W., Bullmore, E.T., 2011. Peripheral biomarkers of cognitive response to dopamine receptor agonist treatment. Psychopharmacology (Berlin) 214 (4), 779–789.

Fink, K.B., Göthert, M., 2007. 5-HT receptor regulation of neurotransmitter release. Pharmacol. Rev. 59 (4), 360–417.

Flagel, S.B., Watson, S.J., Akil, H., Robinson, T.E., 2008. Individual differences in the attribution of incentive salience to a reward-related cue: influence on cocaine sensitization. Behav. Brain Res. 186 (1), 48–56.

Foley, A.G., Murphy, K.J., Hirst, W.D., Gallagher, H.C., Hagan, J.J., Upton, N., et al., 2004. The 5-HT$_6$ receptor antagonist SB-271046 reverses scopolamine-disrupted consolidation of a passive avoidance task and ameliorates spatial task deficits in aged rats. Neuropsychopharmacology 9 (1), 93–100.

Fone, K.C., 2006. Selective 5-HT$_6$ compounds as a novel approach to the treatment of Alzheimer disease. J. Pharmacol. Sci. 101 (Suppl. 1), 53.

Fournet, V., de Lavilléon, G., Schweitzer, A., Giros, B., Andrieux, A., Martres, M.P., 2012. Both chronic treatments by epothilone D and fluoxetine increase the short-term memory and differentially alter the mood-status of STOP/MAP6 KO mice. J. Neurochem. 123 (6), 982–996.

Francis, P.T., Ramırez, M.J., Lai, M.K., 2010. Neurochemical basis for symptomatic treatment of Alzheimer's disease. Neuropharmacology 59, 221–229.

Frauenknecht, K., Katzav, A., Grimm, C., Chapman, J., Sommer, C.J., 2013. Neurological impairment in experimental antiphospholipid syndrome is associated with increased ligand binding to hippocampal and cortical serotonergic 5-HT$_{1A}$ receptors. Immunobiology 218 (4), 517–526.

Freret, T., Paizanis, E., Beaudet, G., Gusmao-Montaigne, A., Nee, G., Dauphin, F., 2013. Modulation of 5-HT$_7$ receptor: effect on object recognition performances in mice. Psychopharmacology (Berlin) [Epub ahead of print] PMID: 23995300.

Gacsályi, I., Nagy, K., Pallagi, K., Lévay, G., Hársing Jr, L.G., Móricz, K., et al., 2013. Egis-11150: a candidate antipsychotic compound with procognitive efficacy in rodents. Neuropharmacology 64, 254–256.

Gallistel, C.R., 2009. The importance of proving the null. Psychol. Rev. 116 (2), 439–453.

Garcia-Alloza, M., Hirst, W.D., Chen, C.P., Lasheras, B., Francis, P.T., Ramírez, M.J., 2004. Differential involvement of 5-HT$_{1B/1D}$ and 5-HT$_6$ receptors in cognitive and non-cognitive symptoms in Alzheimer's disease. Neuropsychopharmacology 29, 410–416.

Gasbarri, A., Cifariello, A., Pompili, A., Meneses, A., 2008. Effect of 5-HT$_7$ antagonist SB-269970 in the modulation of working and reference memory in the rat. Behav. Brain Res. 195 (1), 164–170.

Geldenhuys, W.J., Van der Schyf, C.J., 2009. The serotonin 5-HT$_6$ receptor: a viable drug target for treating cognitive deficits in Alzheimer's disease. Expert Rev. Neurother. 9, 1073–1085.

Gellynck, E., Heyninck, K., Andressen, K.W., Haegeman, G., Levy, F.O., Vanhoenacker, P., et al., 2013. The serotonin 5-HT$_7$ receptors: two decades of research. Exp. Brain Res. [Epub ahead of print] PMID: 24042216.

Gerlai, R., 2001. Behavioral tests of hippocampal function: simple paradigms complex problems. Behav. Brain Res. 125 (1–2), 269–277.

Gobert, A., Lejeune, F., Rivet, J.M., Audinot, V., Newman-Tancredi, A., Millan, M.J., 1995. Modulation of the activity of central serotoninergic neurons by novel serotonin1A receptor agonists and antagonists: a comparison to adrenergic and dopaminergic neurons in rats. J. Pharmacol. Exp. Ther. 273 (3), 1032–1046.

Goghari, V.M., Smith, G.N., Honer, W.G., Kopala, L.C., Thornton, A.E., Su, W., et al., 2013. Effects of eight weeks of atypical antipsychotic treatment on middle frontal thickness in drug-naïve first-episode psychosis patients. Schizophr. Res.pii: S0920-9964(13)00325-3.

Golightly, K.L., Lloyd, J.A., Hobson, J.E., Gallagher, P., Mercer, G., Young, A.H., 2001. Acute tryptophan depletion in schizophrenia. Psychol. Med. 31 (1), 75–84.

Gong, P., Zheng, Z., Chi, W., Lei, X., Wu, X., Chen, D., et al., 2012. An association study of the genetic polymorphisms in 13 neural plasticity-related genes with semantic and episodic memories. J. Mol. Neurosci. 46 (2), 352–361.

Gonzalez, R., Chávez-Pascacio, K., Meneses, A., 2013. Role of 5-HT$_{5A}$ receptors in the consolidation of memory. Behav. Brain Res. 252C, 246−251.

Goodfellow, N.M., Bailey, C.D., Lambe, E.K., 2012. The native serotonin 5-HT$_{5A}$ receptor: electrophysiological characterization in rodent cortex and 5-HT$_{1A}$-mediated compensatory plasticity in the knock-out mouse. J. Neurosci. 32 (17), 5804−5809.

Gravius, A., Laszy, J., Pietraszek, M., Sághy, K., Nagel, J., Chambon, C., et al., 2011. Effects of 5-HT$_6$ antagonists, Ro-4368554 and SB-258585, in tests used for the detection of cognitive enhancement and antipsychotic-like activity. Behav. Pharmacol. 22 (2), 122−135.

Guam, Z., Giustetto, M., Lomvardas, S., Kim, J.H., Miniaci, M.C., Schwartz, J.H., et al., 2002. Integration of long-term-memory-related synaptic plasticity involves bidirectional regulation of gene expression and chromatin structure. Cell 111 (4), 483−493.

Haahr, M.E., Fisher, P., Holst, K., Madsen, K., Jensen, C.G., Marner, L., et al., 2012. The 5-HT$_4$ receptor levels in hippocampus correlates inversely with memory test performance in humans. Hum. Brain Map. Available from: http://dx.doi.org/10.1002/hbm.22123. [Epub ahead of print].

Haider, S., Khaliq, S., Tabassum, S., Haleem, D.J., 2012. Role of somatodendritic and postsynaptic 5-HT$_{1A}$ receptors on learning and memory functions in rats. Neurochem. Res. 37 (10), 2161−2166.

Hajjo, R., Setola, V., Roth, B.L., Tropsha, A., 2012. Chemocentric informatics approach to drug discovery: identification and experimental validation of selective estrogen receptor modulators as ligands of 5-hydroxytryptamine-6 receptors and as potential cognition enhancers. J. Med. Chem. 55 (12), 5704−5719.

Hampel, H., Prvulovic, D., Teipel, S.J., Bokde, A.L., 2011. Recent developments of functional magnetic resonance imaging research for drug development in Alzheimer's disease. Prog. Neurobiol. 95 (4), 570−578.

Hanks, J.B., González-Maeso, J., 2013. Animal models of serotonergic psychedelics. ACS Chem. Neurosci. 4 (1), 33−42.

Hannon, J., Hoyer, D., 2008. Molecular biology of 5-HT receptors. Behav. Brain Res. 195 (1), 198−213.

Harmer, C.J., Cowen, P.J., 2013. 'It's the way that you look at it'—a cognitive neuropsychological account of SSRI action in depression. Phil. Trans. R. Soc. B 368 (1615), 20120407.

Harvey, B.H., Naciti, C., Brand, L., Stein, D.J., 2003. Endocrine, cognitive and hippocampal/cortical 5HT$_{1A/2A}$ receptor changes evoked by a time-dependent sensitisation (TDS) stress model in rats. Brain Res. 983 (1−2), 97−107.

Hawkins, R.D., 2013. Possible contributions of a novel form of synaptic plasticity in Aplysia to reward, memory, and their dysfunctions in mammalian brain. Learn. Mem. 20 (10), 580−591.

Hermann, A., Küpper, Y., Schmitz, A., Walter, B., Vaitl, D., Hennig, J., et al., 2012. Functional gene polymorphisms in the serotonin system and traumatic life events modulate the neural basis of fear acquisition and extinction. PLoS One 7 (9), e44352.

Hernandez-Lopez, S., Garduño, J., Mihailescu, S., 2013. Nicotinic modulation of serotonergic activity in the dorsal raphe nucleus. Rev. Neurosci. 1−15. Available from: http://dx.doi.org/10.1515/revneuro-2013-0012. Available from: http://dx.doi.org/10.1515/revneuro-2013-0012. [Epub ahead of print] PMID: 24021594.

Hindi Attar, C., Finckh, B., Büchel, C., 2012. The influence of serotonin on fear learning. PLoS One 7 (8), e42397.

Hirst, W.D., Andree, T.H., Aschmies, S., Childers, W.E., Comery, T.A., Dawson, L.A., et al., 2008. Correlating efficacy in rodent cognition models with in vivo 5-hydroxytryptamine1a receptor occupancy by a novel antagonist, (R)-*N*-(2-methyl-(4-indolyl-1-piperazinyl)ethyl)-*N*-(2-pyridinyl)-cyclohexane carboxamide (WAY-101405). J. Pharmacol. Exp. Ther. 325 (1), 134–145.

Holenz, J., Pauwels, P.J., Díaz, J.L., Mercè, R., Codony, X., Buschmann, H., 2006. Medicinal chemistry strategies to 5-HT6 receptor ligands as potential cognitive enhancers and antiobesity agents. Drug Discov. Today 11, 283–299.

Hong, E., Meneses, A., 1996. Systemic injection of p-chloroamphetamine eliminates the effect of the 5-HT$_3$ compounds on learning. Pharmacol. Biochem. Behav. 53 (4), 765–769.

Horisawa, T., Nishikawa, H., Toma, S., Ikeda, A., Horiguchi, M., Ono, M., et al., 2013. The role of 5-HT$_7$ receptor antagonism in the amelioration of MK-801-induced learning and memory deficits by the novel atypical antipsychotic drug lurasidone. Behav. Brain Res. 244, 66–69.

Horner, A.E., Heath, C.J., Hvoslef-Eide, M., Kent, B.A., Kim, C.H., Nilsson, S.R., et al., 2013. The touchscreen operant platform for testing learning and memory in rats and mice. Nat. Protoc. 8 (10), 1961–1968.

Hothersall, J.D., Moffat, C., Connolly, C.N., 2013. Prolonged inhibition of 5-HT$_3$ receptors by palonosetron results from surface receptor inhibition rather than inducing receptor internalization. Br. J. Pharmacol. 169 (6), 1252–1262.

Hoyer, et al., 1994. International union of pharmacology classification of receptors for 5-hydroxytryptamine (Serotonin). Pharmacol. Rev. 46 (2), 157–203.

Huang, M., Horiguchi, M., Felix, A.R., Meltzer, H.Y., 2012. 5-HT$_{1A}$ and 5-HT$_7$ receptors contribute to lurasidone-induced dopamine efflux. Neuroreport 23 (7), 436–440.

Huerta-Rivas, A., Perez-Garcia, G., Gonzalez, C., Meneses, A., 2010. Time-course of 5-HT$_6$ receptor mRNA expression during memory consolidation and amnesia. Neurobiol. Learn. Mem. 93 (1), 99–110.

Ivachtchenko, A.V., Ivanenkov, Y.A., 2012. 5HT6 receptor antagonists: a patent update. Part 1: Sulfonyl derivatives. Expert Opin. Ther. Pat. 22 (8), 917–964.

Ivachtchenko, A.V., Ivanenkov, Y.A., Skorenko, A.V., 2012. 5-HT$_6$ receptor modulators: a patent update. Part 2: Diversity in heterocyclic scaffolds. Expert Opin. Ther. Pat. 22 (10), 1123–1168.

Izquierdo, I., Cunha, C., Medina, J.H., 1990. Endogenous benzodiazepine modulation of memory processes. Neurosci. Biobehav. Rev. 14 (4), 419–424.

Izquierdo, I., Medina, J.H., Izquierdo, L.A., Barros, D.M., de Souza, M.M., Mello e Souza, T., 1998. Short- and long-term memory are differentially regulated by monoaminergic systems in the rat brain. Neurobiol. Learn. Mem. 69 (3), 219–224.

Izquierdo, I., Medina, J.H., Vianna, M.R., Izquierdo, L.A., Barros, D.M., 1999. Separate mechanisms for short- and long-term memory. Behav. Brain Res. 103 (1), 1–11.

Izquierdo, I., Bevilaqua, L.R., Rossato, J.I., Bonini, J.S., Da Silva, W.C., Medina, J.H., et al., 2006. The connection between the hippocampal and the striatal memory systems of the brain: a review of recent findings. Neurotox. Res. 10 (2), 113–121.

Jacobs, B.L., Azmitia, E.C., 1992. Structure and function of the brain serotonin system. Physiol. Rev. 72, 165–229.

Jarome, T.J., Lubin, F.D., 2013. Histone lysine methylation: critical regulator of memory and behavior. Rev. Neurosci. 27, 1–13.

Jen, C.J., Lin, L.C., Wu, F.S., Ho, Y.C., Chen, H.I., 2008. Treadmill exercise enhances passive avoidance learning in rats: the role of down-regulated serotonin system in the limbic system.

Front. Hum. Neurosci.Conference Abstract: 10th International Conference on Cognitive Neuroscience. doi:10.3389/conf.neuro.09.2009.01.285.

Jensen, A.A., Plath, N., Pedersen, M.H., Isberg, V., Krall, J., Wellendorph, P., et al., 2013. Design, synthesis, and pharmacological characterization of N- and O-substituted 5,6,7,8-tetrahy-dro-4H-isoxazolo[4,5-d]azepin-3-ol analogues: novel 5-HT(2A)/5-HT(2C) receptor agonists with pro-cognitive properties. J. Med. Chem. 56 (3), 1211–1227.

Johnson, C.N., Ahmed, M., Miller, N.D., 2008. $5-HT_6$ receptor antagonists: prospects for the treatment of cognitive disorders including dementia. Curr. Opin. Drug Discov. Devel. 11, 642–654.

Jones, T., Moller, M.D., 2011. Implications of hypothalamic–pituitary–adrenal axis functioning in posttraumatic stress disorder. J. Am. Psychiatr. Nurses Assoc. 17 (6), 393–403.

Kandel, E.R., 2001. The molecular biology of memory storage: a dialogue between genes and synapses. Science 294, 1030–1038.

Kandel, E.R., 2012. The molecular biology of memory: cAMP, PKA, CRE, CREB-1, CREB-2, and CPEB. Mol. Brain 5, 14.

Karanges, E.A., Kashem, M.A., Sarker, R., Ahmed, E.U., Ahmed, S., Van Nieuwenhuijzen, P.S., et al., 2013. Hippocampal protein expression is differentially affected by chronic paroxetine treatment in adolescent and adult rats: a possible mechanism of "paradoxical" antidepressant responses in young persons. Front. Pharmacol.. Available from: http://dx.doi.org/10.3389/fphar.2013.00086.

Kassai, F., Schlumberger, C., Kedves, R., Pietraszek, M., Jatzke, C., Lendvai, B., et al., 2012. Effect of $5-HT_{5A}$ antagonists in animal models of schizophrenia, anxiety and depression. Behav. Pharmacol. 23 (4), 397–406.

Kemp, A., Manahan-Vaughan, D., 2005. The 5-hydroxytryptamine4 receptor exhibits frequency-dependent properties in synaptic plasticity and behavioural metaplasticity in the hippocampal CA1 region in vivo. Cereb. Cortex 15 (7), 1037–1043.

Khilnani, G., Khilnani, A.K., 2011. Inverse agonism and its therapeutic significance. Indian J. Pharmacol. 43 (5), 492–501.

Kikuchi, C., Suzuki, H., Hiranuma, T., Koyama, M., 2003. New tetrahydrobenzindoles as potent and selective $5-HT_7$ antagonists with increased in vitro metabolic stability. Bioorg. Med. Chem. Lett. 13 (1), 61–64.

King, M.V., Marsden, C.A., Fone, K.C., 2008. A role for the $5-HT_{1A}$, $5-HT_4$ and $5-HT_6$ receptors in learning and memory. Trends Pharmacol. Sci. 29, 482–492.

Kirkwood, A., 2000. Serotonergic control of developmental plasticity. Proc. Natl. Acad Sci. USA 97 (5), 1951–1952.

Klein, M.T., Teitler, M., 2012. Distribution of $5-ht_{1E}$ receptors in the mammalian brain and cerebral vasculature: an immunohistochemical and pharmacological study. Br. J. Pharmacol. 166 (4), 1290–1302.

Kondo, M., Nakamura, Y., Ishida, Y., Yamada, T., Shimada, S., 2013. The $5-HT_{3A}$ receptor is essential for fear extinction. Learn. Mem. 21 (1), 740–743.

Kuhn, M., Popovic, A., Pezawas, L., 2013. Neuroplasticity and memory formation in major depressive disorder: an imaging genetics perspective on serotonin and BDNF. Restor. Neurol. Neurosci. [Epub ahead of print] PMID: 23603442.

Lamirault, L., Simon, H., 2001. Enhancement of place and object recognition memory in young adult and old rats by RS 67333, a partial agonist of $5-HT_4$ receptors. Neuropharmacology 41 (7), 844–853.

Lelong, V., Lhonneur, L., Dauphin, F., Boulouard, M., 2006. BIMU 1 and RS 67333, two $5-HT_4$ receptor agonists, modulate spontaneous alternation deficits induced by scopolamine in the mouse. Naunyn-Schmiedeberg's Arch. Pharmacol. 367, 621–628.

Leopoldo, M., Lacivita, E., Berardi, F., 2010. Perrone R.5-HT$_7$ receptor modulators: a medicinal chemistry survey of recent patent literature (2004–2009). Expert Opin. Ther. Pat. 20 (6), 739–754.

Leopoldo, M., Lacivita, E., Berardi, F., Perrone, R., Hedlund, P.B., 2011. Serotonin 5-HT$_7$ receptor agents: structure–activity relationships and potential therapeutic applications in central nervous system disorders. Pharmacol. Ther. 129 (2), 120–148.

Lesch, K.P., Waider, J., 2012. Serotonin in the modulation of neural plasticity and networks: implications for neurodevelopmental disorders. Neuron 76 (1), 175–191.

Linder, M.D., Hodges, D.B., Hogan, J.B., Orie, A.F., Corsa, J.A., Bartyen, D.M., et al., 2003. An assessment of the effects of 5-HT$_6$ receptor antagonists in rodent models of learning. J. Pharmacol. Exp. Ther. 307, 682–691.

Liu, G.L., Robichaud, A.J., 2009. 5-HT$_6$ antagonists as potential treatment for cognitive dysfunction. Drug Dev. Res. 70, 145–168.

Lladó-Pelfort, L., Santana, N., Ghisi, V., Artigas, F., Celada, P., 2012. 5-HT$_{1A}$ receptor agonists enhance pyramidal cell firing in prefrontal cortex through a preferential action on GABA interneurons. Cereb. Cortex 22 (7), 1487–1497.

Lopez-Velazquez, M.A., Gutíerrez-Guzmán, E., Cervantes, M., Olvera–Cortez, M.E., 2011. 5-HT$_{2C}$ receptors in learning, doi:10.10007/978-1-60761-941-3_24 .In: Di Giovanni, G., Esposito, E., Di Matteo, V. (Eds.), 5-HT2C Receptors in the Pathology of CSN Disease. Springer Science + Business Media, LLC, Totowa, NJ., pp. 461–507.

Lorke, D.E., Lu, G., Cho, E., Yew, D.T., 2006. Serotonin 5-HT$_{2A}$ and 5-HT$_6$ receptors in the prefrontal cortex of Alzheimer and normal aging patients. BMC Neurosci. 7, 36.

Lucaites, V.L., Krushinsk, J.H., Schaus, J.M., Audia, J.E., Nelson, D.L., 2005. [^{3}H]LY334370, a novel radioligand for the 5-HT$_{1F}$ receptor. II. Autoradiographic localization in rat, guinea pig, monkey and human brain. Naunyn-Schmiedeberg's Arch. Pharmacol. 371, 178–184.

Ly, S., Pishdari, B., Lok, L.L., Hajos, M., Kocsis, B., 2013. Activation of 5-HT$_6$ receptors modulate sleep–wake activity and hippocampal Theta oscillation. ACS Chem. 4, 191–199.

Lynch, G., Rex, C.S., Chen, L.Y., Gall, C.M., 2008. The substrates of memory: defects, treatments, and enhancement. Eur. J. Pharmacol. 585 (1), 2–13.

Lynch, M.A., 2004. Long-term potentiation and memory. Physiol. Rev. 84, 87–136.

Lyseng-Williamson, K., 2013. Clozapine: a guide to its use in patients with schizophrenia who are unresponsive to or intolerant of other antipsychotic agents. Drug Ther. Perspect. 29, 161–165.

Maher-Edwards, G., Zvartau-Hind, M., Hunter, A.J., Gold, M., Hopton, G., Jacobs, G., et al., 2009. Controlled phase II study of a 5-HT$_6$ receptor antagonist, SB-742457. Curr. Alzheimer Res. 7 (5), 374–385.

Maillet, M., Robert, S.J., Lezoualc'h, F., 2004. New insights into serotonin 5-HT$_4$ receptors: a novel therapeutic target for Alzheimer's disease? Curr. Alzheimer Res. 1, 79–85.

Markou, A., Salamone, J.D., Bussey, T.J., Mar, A.C., Brunner, D., Gilmour, G., et al., 2013. Measuring reinforcement learning and motivation constructs in experimental animals: Relevance to the negative symptoms of schizophrenia. Neurosci. Biobehav. Rev. 37 (9B), 2149–2165.

Malá, H., Arnberg, K., Chu, D., Nedergaard, S.K., Witmer, J., Mogensen, J., 2013. Only repeated administration of the serotonergic agonist 8-OH-DPAT improves place learning of rats subjected to fimbria-fornix transection. Pharmacol. Biochem. Behav. 109, 50–58.

Manuel-Apolinar, L., Meneses, A., 2004. 8-OH-DPAT facilitated memory consolidation and increased hippocampal and cortical cAMP production. Behav. Brain Res. 148 (1–2), 179–184.

Manuel-Apolinar, L., Rocha, L., Pascoe, D., Castillo, E., Castillo, C., Meneses, A., 2005. Modifications of 5-HT$_4$ receptor expression in rat brain during memory consolidation. Brain Res. 1042 (1), 73–81.

Mar, A.C., Horner, A.E., Nilsson, S.R., Alsiö, J., Kent, B.A., Kim, C.H., et al., 2013. The touchscreen operant platform for assessing executive function in rats and mice. Nat. Protoc. 8 (10), 1985–2005.

Marchetti, E., Jacquet, M., Escoffier, G., Miglioratti, M., Dumuis, A., Bockaert, J., et al., 2011. Enhancement of reference memory in aged rats by specific activation of 5-HT$_4$ receptors using an olfactory associative discrimination task. Brain Res. 1405, 49–56.

Marcos, B., Gil-Bea, F.J., Hirst, W.D., Garcia-Alloza, M., Ramírez, M.J., 2006. Lack of localization of 5-HT$_6$ receptors on cholinergic neurons: implication of multiple neurotransmitter systems in 5-HT$_6$ receptor-mediated acetylcholine release. Eur. J. Neurosci. 24, 299–306.

Marcos, B., Garcia-Alloza, M., Gil-Bea, F.J., Chuang, T.T., Francis, P.T., Chen, C.P., et al., 2008. Involvement of an altered 5-HT$_6$ receptor function in behavioral symptoms of Alzheimer's disease. J. Alzheimers Dis. 14, 43–50.

Marcos, B., Cabero, M., Solas, M., Aisa, B., Ramirez, M.J., 2010. Signaling pathways associated with 5-HT$_6$ receptors: relevance for cognitive effects. Int. J. Neuropsychopharmacol. 9, 1–10.

Marin P., Meffre J., Chaumont-Dubel S., La Cour C.L., Loiseau F., Watson D.J.G., et al., 5-HT$_6$ Receptors Disrupt Cognition by Recruiting mTOR: Relevance to Schizophrenia. Serotonin Club meeting, Abstract p: 58, 2012a, Montpellier, France.

Marin, P., Becamel, C., Dumuis, A., Bockaert, J., 2012b. 5-HT receptor-associated protein networks: new targets for drug discovery in psychiatric disorders? Curr. Drug Targets 13 (1), 28–52.

Marshall, J.F., O'Dell, S.J., 2012. Methamphetamine influences on brain and behavior: unsafe at any speed? Trends Neurosci. 35 (9), 536–545.

Martin, C., Sibson, N.R., 2008. Pharmacological MRI in animal models: a useful tool for 5-HT research? Neuropharmacology 55, 1038–1047.

Mathur, B.N., Lovinger, D.M., 2012. Serotonergic action on dorsal striatal function. Parkinsonism Relat. Disord. 18 (Suppl. 1), S129–S131.

Matthys, A., Haegeman, G., Van Craenenbroeck, K., Vanhoenacker, P., 2011. Role of the 5-HT$_7$ receptor in the central nervous system: from current status to future perspectives. Mol. Neurobiol. 43 (3), 228–253.

McIntosh, A.L., Ballard, T.M., Steward, L.J., Moran, P.M., Fone, K.C., 2013. The atypical antipsychotic risperidone reverses the recognition memory deficits induced by post-weaning social isolation in rats. Psychopharmacology (Berlin) 228 (1), 31–42.

McLoughlin, D.J., Strange, P.G., 2000. Mechanisms of agonism and inverse agonism at serotonin 5-HT$_{1A}$ receptors. J. Neurochem. 74 (1), 347–357.

Meltzer, H.Y., 2012. Clozapine: balancing safety with superior antipsychotic efficacy. Clin. Schizophr. Relat. Psychoses 6 (3), 134–144.

Meltzer, H.Y., Massey, B.W., Horiguchi, M., 2012. Serotonin receptors as targets for drugs useful to treat psychosis and cognitive impairment in schizophrenia. Curr. Pharm. Biotechnol. 13 (8), 1572–1586.

Meneses, A., 1999. 5-HT system and cognition. Neurosci. Biobehav. Rev. 23, 1111–1125.

Meneses, A., 2001. Could the 5-HT$_{1B}$ receptor inverse agonism affects learning consolidation? Neurosci. Biobehav. Rev. 25 (2), 193–201.

Meneses, A., 2002. Involvement of 5-HT$_{2A/2B/2C}$ receptors on memory formation: simple agonism, antagonism, or inverse agonism? Cell. Mol. Neurobiol. 22 (5–6), 675–688.

Meneses, A., 2003. Pharmacological analysis of an associative learning task: 5-HT$_1$ to 5-HT$_7$ receptor subtypes function on a Pavlovian/instrumental autoshaped memory. Learn. Mem. 10, 363–372.

Meneses, A., 2004. Effects of the 5-HT$_7$ receptor antagonists SB-269970 and DR 4004 in autoshaping Pavlovian/instrumental learning task. Behav. Brain Res. 155 (2), 275–282.

Meneses, A., 2007a. Stimulation of 5-HT$_{1A}$, 5-HT$_{1B}$, 5-HT$_{2A/2C}$, 5-HT$_3$ and 5-HT$_4$ receptors or 5-HT uptake inhibition: short- and long-term memory. Behav. Brain Res. 184 (1), 81−90.

Meneses, A., 2007b. Do serotonin$_{1-7}$ receptors modulate short and long-term memory? Neurobiol. Learn. Mem. 87 (4), 561−572.

Meneses, A., 2013. 5-HT systems: emergent targets for memory formation and memory alterations. Rev. Neurosci. 24 (6), 629−664.

Meneses, A., 2014. Neurotransmitters and memory: cholinergic, glutamatergic, gabaergic, dopaminergic, serotonergic, signaling, and memory. In: Meneses, A. (Ed.), Identification of Neural Markers Accompanying Memory. Elsevier, San Diego, CA, pp. 5−45.

Meneses, A., Hong, E., 1997. Effects of 5-HT$_4$ receptor agonist and antagonist in learning. Pharmacol. Biochem. Behav. 56, 347−351.

Meneses, A., Liy-Salmeron, G., 2012. Serotonin and emotion, learning and memory. Rev. Neurosci. 23, 443−453.

Meneses, A., Perez-Garcia, G., 2007. 5-HT$_{1A}$ receptors and memory. Neurosci. Biobehav. Rev. 31 (5), 705−727.

Meneses, A., Manuel-Apolinar, L., Rocha, L., Castillo, E., Castillo, C., 2004a. Expression of the 5-HT receptors in rat brain during memory consolidation. Behav. Brain Res. 152 (2), 425−436.

Meneses, A., Manuel-Apolinar, L., Castillo, C., Castillo, E., 2004b. Memory consolidation and amnesia modify 5-HT$_6$ receptors expression in rat brain: an autoradiographic study. Behav. Brain Res. 178, 53−61.

Meneses, A., Perez-Garcia, G., Liy-Salmeron, G., Ponce-Lopez, T., Tellez, R., Flores-Galvez, D., 2009. Associative learning, memory and serotonin: a neurobiological and pharmacological analysis, ISBN: 978-81-7895-383-0. In: Rocha Arrieta, L.L., Granados-Soto, V. (Eds.), Models of Neuropharmacology. Transworld Research Network, Trivandrum, Kerala, India, pp. 169−182.

Meneses, A., Perez-Garcia, G., Ponce-Lopez, T., Castillo, C., 2011a. 5-HT$_6$ receptor memory and amnesia: behavioral pharmacology—learning and memory processes. Int. Rev. Neurobiol. 96, 27−47.

Meneses, A., Perez-Garcia, G., Ponce-Lopez, T., Tellez, R., Castillo, C., 2011b. Serotonin transporter and memory. Neuropharmacology 61 (3), 355−363.

Meyer, J.H., 2012. Neuroimaging markers of cellular function in major depressive disorder: implications for therapeutics, personalized medicine, and prevention. Clin. Pharmacol. Ther. 91 (2), 201−214.

Millan, M.J., 2011. MicroRNA in the regulation and expression of serotonergic transmission in the brain and other tissues. Curr. Opin. Pharmacol. 11 (19), 11−22.

Millan, M.J., Gobert, A., Roux, S., Porsolt, R., Meneses, A., Carli, M., et al., 2004. The serotonin1A receptor partial agonist S15535 [4-(benzodioxan-5-yl)1-(indan-2-yl)piperazine] enhances cholinergic transmission and cognitive function in rodents: a combined neurochemical and behavioral analysis. J. Pharmacol. Exp. Ther. 311 (1), 190−203.

Millan, M.J., Agid, Y., Brüne, M., Bullmore, E.T., Carter, C.S., Clayton, N.S., et al., 2012. Cognitive dysfunction in psychiatric disorders: characteristics, causes and the quest for improved therapy. Nat. Rev. Drug Discov. 11 (2), 141−168.

Mitchell, E.S., Neumaier, J.F., 2005. 5-HT$_6$ receptors: a novel target for cognitive enhancement. Pharmacol. Ther. 108, 320−333.

Mitchell, E.S., Sexton, T., Neumaier, J.F., 2007. Increased expression of 5-HT$_6$ receptors in the rat dorsomedial striatum impairs instrumental learning. Neuropsychopharmacology 32, 1520−1530.

Mnie-Filali, O., Lambás-Señas, L., Zimmer, L., Haddjeri, N., 2007. 5-HT$_7$ receptor antagonists as a new class of antidepressants. Drug News Perspect. 20 (10), 613−618.

Molinuevo, J.L., Sánchez-Valle, R., Lladó, A., Fortea, J., Bartrés-Faz, D., Rami, L., 2012. Identifying earlier Alzheimer's disease: insights from the preclinical and prodromal phases. Neurodegener. Dis. 10 (1–4), 58–160.

Monje, F.J., Divisch, I., Demit, M., Lubec, G., Pollak, D.D., 2013. Flotillin-1 is an evolutionary-conserved memory-related protein up-regulated in implicit and explicit learning paradigms. Ann. Med. [Epub ahead of print] PMID: 23631399.

Morris, R., 1984. Developments of a water-maze procedure for studying spatial learning in the rat. J. Neurosci. Methods 11, 47–60.

Myhrer, T., 2003. Neurotransmitter systems involved in learning and memory in the rat: a meta-analysis based on studies of four behavioral tasks. Brain Res. Brain Res. Rev. 41 (2–3), 268–287.

Na, C.H., Jones, D.R., Yang, Y., Wang, X., Xu, Y., Peng, J., 2012. Synaptic protein ubiquitination in rat brain revealed by antibody-based ubiquitome analysis. J. Proteome Res. 11 (9), 4722–4732.

Navailles, S., Lagière, M., Guthriè, M., De Deurwaerdère, P., 2013. Serotonin2C receptor constitutive activity: in vivo direct and indirect evidence and functional significance. Cent. Nerv. Syst. Agents Med. Chem. [Epub ahead of print] PMID: 23441866.

Newman, A.S., Batis, N., Grafton, G., Caputo, F., Brady, C.A., Lambert, J.J., et al., 2013. 5-Chloroindole: a potent allosteric modulator of the $5\text{-}HT_3$ receptor. Br. J. Pharmacol. 169 (6), 1228–1238.

Newman-Tancredi, A., Kleven, M.S., 2011. Comparative pharmacology of antipsychotics possessing combined dopamine D2 and serotonin $5\text{-}HT_{1A}$ receptor properties. Psychopharmacology (Berlin) 216 (4), 451–473.

Nikiforuk, A., Popik, P., 2013. Amisulpride promotes cognitive flexibility in rats: the role of $5\text{-}HT_7$ receptors. Behav. Brain Res. 248, 136–140.

Nikiforuk, A., Fijaø, K., Potasiewicz, A., Popik, P., Kos, T., 2013. The 5-hydroxytryptamine (serotonin) receptor 6 agonist EMD 386088 ameliorates ketamine-induced deficits in attentional set shifting and novel object recognition, but not in the prepulse inhibition in rats. J. Psychopharmacol. 27 (5), 469–476.

Nonkes, L.J., Maes, J.H., Homberg, J.R., 2013. Improved cognitive flexibility in serotonin transporter knockout rats is unchanged following chronic cocaine self-administration. Addict. Biol. 18 (3), 434–440.

Nonkes, L.J., van de Vondervoort, I.I., Homberg, J.R., 2014. The attribution of incentive salience to an appetitive conditioned cue is not affected by knockout of the serotonin transporter in rats. Behav. Brain Res. 259, 268–273.

Nordquist, N., Oreland, L., 2010. Serotonin, genetic variability, behaviour, and psychiatric disorders—a review. Ups J. Med. Sci. 115 (1), 2–10.

Noristani, H.N., Verkhratsky, A., Rodríguez, J.J., 2012. High tryptophan diet reduces CA1 intra-neuronal β-amyloid in the triple transgenic mouse model of Alzheimer's disease. Aging Cell 11 (5), 810–822.

Ögren, S.O., Eriksson, T.M., Elvander-Tottie, E., D'Addario, C., Ekström, J.C., Svenningsson, P., et al., 2008. The role of $5\text{-}HT_{1A}$ receptors in learning and memory. Behav. Brain Res. 195, 54–77.

Olivier, J.D., Vinkers, C.H., Olivier, B., 2013. The role of the serotonergic and GABA system in translational approaches in drug discovery for anxiety disorders. Front. Pharmacol. 11 (4), 74.

Olshavsky1, M.E., Jones, C.E., Lee, H.J., Monfils, M.H., 2013. Sign-tracking phenotypes during appetitive learning influence open field behavior and maintenance of conditioned fear. Soc. Neurosci. Abstract 93.06/HHH22.

Park, S.M., Williams, C.L., 2012. Contribution of serotonin type 3 receptors in the successful extinction of cued or contextual fear conditioned responses: interactions with GABAergic signaling. Rev. Neurosci. 23 (5–6), 555–569.

Parrott, A.C., 2013a. MDMA, serotonergic neurotoxicity, and the diverse functional deficits of recreational 'Ecstasy' users. Neurosci. Biobehav. Rev. 37 (8), 1466–1484.

Parrott, A.C., 2013b. Human psychobiology of MDMA or 'Ecstasy': an overview of 25 years of empirical research. Hum. Psychopharmacol. 28 (4), 289–307.

Pattij, T., 2002. 5-HT$_{1A}$ Receptor Knockout Mice and Anxiety: Behavioral and Physiological Studies (Ph.D. thesis). Universiteit Utrecht, The Netherlands.

Peele, D.B., Vincent, A., 1989. Strategies for assessing learning and memory, 1978–1987: a comparison of behavioral toxicology, psychopharmacology, and neurobiology. Neurosci. Biobehav. Rev. 13 (4), 317–322.

Pennanen, L., van der Hart, M., Yu, L., Tecott, L.H., 2013. Impact of serotonin (5-HT)$_{2C}$ receptors on executive control processes. Neuropsychopharmacology 38 (6), 957–967.

Perez-Garcia, G., Meneses, A., 2005a. Oral administration of the 5-HT$_6$ receptor antagonists SB-357134 and SB-399885 improves memory formation in an autoshaping learning task. Pharmacol. Biochem. Behav. 81 (3), 673–682.

Perez-Garcia, G., Meneses, A., 2008a. Memory formation, amnesia, improved memory and reversed amnesia: 5-HT role. Behav. Brain Res. 195, 17–29.

Perez-Garcia, G., Meneses, A., 2008b. Ex-vivo study of 5-HT$_{1A}$ and 5-HT$_7$ receptor agonists and antagonists on cAMP accumulation during memory formation and amnesia. Behav. Brain Res. 195, 139–146.

Perez-Garcia, G., Meneses, A., 2009. Memory time-course: mRNA 5-HT$_{1A}$ and 5-HT$_7$ receptors. Behav. Brain Res. 202, 102–113.

Perez-Garcia, G., Gonzalez-Espinosa, C., Meneses, A., 2006. An mRNA expression analysis of stimulation and blockade of 5-HT$_7$ receptors during memory consolidation. Behav. Brain Res. 169, 83–92.

Perez-Garcia, G.S., Meneses, A., 2005b. Effects of the potential 5-HT$_7$ receptor agonist AS 19 in an autoshaping learning task. Behav. Brain Res. 163 (1), 136–140.

Ponce-Lopez, T., Meneses, A., 2008 Effects of the 5-HT1F receptor agonist LY344864 during memory formation, amnesia and camp. 38th Society for Neuroscience Meeting, Washington DC, USA, November 15–19, 2008. Program No. XXX.XX. 2008 Neuroscience Meeting Planner. Society for Neuroscience, Washington, DC. Abstract no. 290.12/RR63.

Porter, R.J., Lunn, B.S., O'Brien, J.T., 2003. Effects of acute tryptophan depletion on cognitive function in Alzheimer's disease and in the healthy elderly. Psychol. Med. 33 (1), 41–49.

Preethi, J., Singh, H.K., Charles, P.D., Rajan, K.E., 2012. Participation of microRNA 124-CREB pathway: a parallel memory enhancing mechanism of standardised extract of Bacopa monniera (BESEB CDRI-08). Neurochem. Res. 37 (10), 2167–2177.

Puig, M.V., Gulledge, A.T., 2011. Serotonin and prefrontal cortex function: neurons, networks, and circuits. Mol. Neurobiol. 44 (3), 449–464.

Rahn, E.J., Guzman-Karlsson, M.C., David Sweatt, J., 2013. Cellular, molecular, and epigenetic mechanisms in non-associative conditioning: implications for pain and memory. Neurobiol. Learn. Mem. pii: S1074-7427(13)00100-7. 10.1016/j.nlm.2013.06.008. [Epub ahead of print].

Rajan, K.E., Singh, H.K., Parkavi, A., Charles, P.D., 2011. Attenuation of 1-(m-chlorophenyl)-biguanide induced hippocampus-dependent memory impairment by a standardised extract of Bacopa monniera (BESEB CDRI-08). Neurochem. Res. 36 (11), 2136–2144.

Rajasethupathy, P., Antonov, I., Sheridan, R., Frey, S., Sander, C., Tuschl, T., et al., 2012. A role for neuronal piRNAs in the epigenetic control of memory-related synaptic plasticity. Cell 149 (3), 693−707.

Ramírez, M.J., 2013. 5-HT$_6$ receptors and Alzheimer's disease. Alzheimers Res. Ther. 5 (2), 15.

Reggio, P.H., 2003. Pharmacophores for ligand recognition and activation/inactivation of the cannabinoid receptors. Curr. Pharm. Des. 9 (20), 1607−1633.

Reid, M., Carlyle, I., Caulfield, W.L., Clarkson, T.R., Cusick, F., Epemolu, O., et al., 2010. The discovery and SAR of indoline-3-carboxamides—A new series of 5-HT$_6$ antagonists. Bioorg. Med. Chem. Lett. 20 (12), 3713−3716.

Renner, U., Zeug, A., Woehler, A., Niebert, M., Dityatev, A., Dityateva, G., et al., 2012. Heterodimerization of serotonin receptors 5-HT$_{1A}$ and 5-HT$_7$ differentially regulates receptor signalling and trafficking. J. Cell. Sci. 125 (Pt 10), 2486−2499.

Renoir, T., Pang, T.Y., Lanfumey, L., 2012. Drug withdrawal-induced depression: serotonergic and plasticity changes in animal models. Neurosci. Biobehav. Rev. 36 (1), 696−726.

Richetto, J., Calabrese, F., Meyer, U., Riva, M.A., 2013. Prenatal versus postnatal maternal factors in the development of infection-induced working memory impairments in mice. Brain Behav. Immun. pii: S0889-1591(13)00240-7. 10.1016/j.bbi.2013.07.006. [Epub ahead of print] PMID: 23876745.

Roberts, A.J., Hedlund, P.B., 2012. The 5-HT$_7$ receptor in learning and memory. Hippocampus 22 (4), 762−771.

Rodríguez, J.J., Noristani, H.N., Verkhratsky, A., 2012. The serotonergic system in ageing and Alzheimer's disease. Prog. Neurobiol. 99, 15−41.

Rogers, D.C., Hagan, J.J., 2001. 5-HT$_6$ receptor antagonists enhance retention of a water maze task in the rat. Psychopharmacology (Berlin) 158 (2), 114−119.

Roman, F.S., Marchetti, E., 1998. Involvement of 5-HT receptors in learning and memory. IDrugs 1 (1), 109−121.

Romero, G., Sánchez, E., Pujol, M., Pérez, P., Codony, X., Holenz, J., et al., 2006. Efficacy of selective 5-HT$_6$ receptor ligands determined by monitoring 5-HT6 receptor-mediated cAMP signaling pathways. Br. J. Pharmacol. 148 (8), 1133−1143.

Rossi-Arnaud, C., Ammassari-Teule, M., 1998. What do comparative studies of inbred mice add to current investigations on the neural basis of spatial behaviors? Exp. Brain Res. 123 (1-2), 36−44.

Rossé, G., Schaffhauser, H., 2010. 5-HT$_6$ receptor antagonists as potential therapeutics for cognitive impairment. Curr. Top. Med. Chem. 10, 207−221.

Roth, B.L., Hanizavareh, S.M., Blum, A.E., 2004. Serotonin receptors represent highly favorable molecular targets for cognitive enhancement in schizophrenia and other disorders. Psychopharmacology (Berlin) 174, 17−24.

Ruiz, N.V., Oranias, G.O., 2010. Patents. In: Borsini, F. (Ed.), International Review of Neurobiology. 5-HT6 Receptors, Part I. Elsevier, Academic Press, Oxford, pp. 36−66.

Russell, M.G., Dias, R., 2002. Memories are made of this (perhaps): a review of serotonin 5-HT$_6$ receptor ligands and their biological functions. Curr. Top. Med. Chem. 2 (6), 643−654.

Salminen, L.E., Schofield, P.R., Pierce, K.D., Lane, E.M., Heaps, J.M., Bolzenius, J.D., et al., 2013. Triallelic relationships between the serotonin transporter polymorphism and cognition among healthy older adults. Int. J. Neurosci. [Epub ahead of print] PMID: 24044728.

Sarter, M., Stephens, D.N., 1988. Beta-carbolines as tools in memory research: animal data and speculations. Psychopharmacol. Ser. 6, 230−245.

Saulin, A., Savli, M., Lanzenberger, R., 2012. Serotonin and molecular neuroimaging in humans using PET. Amino Acids 42 (6), 2039−2057.

Sawyer, J., Eaves, E.L., Heyser, C.J., Maswood, S., 2012. Tropisetron, a 5-HT$_3$ receptor antagonist, enhances object exploration in intact female rats. Behav. Pharmacol. 23 (8), 806−809.

Schechter, L.E., Smith, D.L., Zhang, G.M., Li, P., Lin, Q., Rosenzweig-Lipson, S., et al., 2004. WAY-466: pharmacological profile of a novel and selective 5-HT$_6$ agonist. Int. J. Neuropsychopharmacol. 7 (Suppl. 1), S291.

Schechter, L.E., Smith, D.L., Rosenzweig-Lipson, S., Sukoff, S.J., Dawson, L.A., Marquis, K., et al., 2005. Lecozotan (SRA-333): a selective Serotonin1A receptor antagonist that enhances the stimulated release of glutamate and acetylcholine in the hippocampus and possesses cognitive-enhancing properties. J. Exp. Pharmacol. Ther. 314, 1274−1289.

Schiller, L., Jäjkel, M., Kretzschmar, M., Brust, P., Oehler, J., 2003. Autoradiographic analyses of 5-HT$_{1A}$ and 5-HT$_{2A}$ receptors after social isolation in mice. Brain Res. 980, 169−178.

Schmitt, J.A., Wingen, M., Ramaekers, J.G., Evers, E.A., Riedel, W.J., 2006. Serotonin and human cognitive performance. Curr. Pharm. Des. 12 (20), 2473−2486.

Sharp, T., Bramwell, S.R., Hjorth, S., Grahame-Smith, D.G., 1989. Pharmacological characterization of 8-OH-DPAT-induced inhibition of rat hippocampal 5-HT release in vivo as measured by microdialysis. Br. J. Pharmacol. 98 (3), 989−997.

Shelton, A.L., Marchette, S.A., Furman, A.J., 2014. A mechanistic approach to individual differences in spatial learning, memory, and navigation. Psychol. Learn. Motivation Adv. Res. Theory 59, 223−259.

Shen, F., Smith, J.A., Chang, R., Bourdet, D.L., Tsuruda, P.R., Obedencio, G.P., et al., 2011. 5-HT$_4$ receptor agonist mediated enhancement of cognitive function in vivo and amyloid precursor protein processing in vitro: a pharmacodynamic and pharmacokinetic assessment. Neuropharmacology 61 (1−2), 69−79.

Shimizu, S., Mizuguchi, Y., Ohno, Y., 2013. Improving the treatment of Schizophrenia: role of 5-HT receptors in modulating cognitive and extrapyramidal motor functions. CNS Neurol. Disord. Drug. Targets [Epub ahead of print] PMID: 23844689 [PubMed].

Singh, C., Bortolato, M., Bali, N., Godar, S.C., Scott, A.L., Chen, K., et al., 2013. Cognitive abnormalities and hippocampal alterations in monoamine oxidase A and B knockout mice. Proc. Natl. Acad. Sci. USA [Epub ahead of print] PMID: 23858446.

Smith, C., Rahman, T., Toohey, N., Mazurkiewicz, J., Herrick-Davis, K., Teitler, M., 2006. Risperidone irreversibly binds to and inactivates the h5-HT$_7$ serotonin receptor. Mol. Pharmacol. 70 (4), 1264−1270.

Smith, C., Toohey, N., Knight, J.A., Klein, M.T., Teitler, M., 2011. Risperidone-induced inactivation and clozapine-induced reactivation of rat cortical astrocyte 5-hydroxytryptamine$_7$ receptors: evidence for in situ G protein-coupled receptor homodimer protomer cross-talk. Mol. Pharmacol. 79 (2), 318−325.

Sodhi, M.S., Sanders-Bush, E., 2004. Serotonin and brain development. Int. Rev. Neurobiol. 59, 111−174.

Steinbush, H.W.M., 1984. Serotonin-immunoreactive neurons and their projections in the CNS. In: Bjorklund, A., Hokfelt, T., Kuhar, M.J. (Eds.), Handbook of Chemical Neuroanatomy. Classical Neurotransmitters and Transmitters Receptors in the CNS, Part III, vol. 3. Elsevier, Amsterdam, pp. 68−125.

Sumiyoshi, T., Bubenikova-Valesova, V., Horacek, J., Bert, B., 2008. Serotonin1A receptors in the pathophysiology of schizophrenia: development of novel cognition-enhancing therapeutics. Adv. Ther. 25 (10), 1037−1056.

Sumiyoshi, T., Higuchi, Y., Uehara, T., 2013. Neural basis for the ability of atypical antipsychotic drugs to improve cognition in schizophrenia. Front. Behav. Neurosci. 7, 140.

Tarazi, F.I., Riva, M.A., 2013. The preclinical profile of lurasidone: clinical relevance for the treatment of schizophrenia. Expert Opin. Drug Discov. [Epub ahead of print] PMID: 23837554.

Tellez, R., Rocha, L., Castillo, C., Meneses, A., 2010. Autoradiographic study of serotonin transporter during memory formation. Behav. Brain Res. 212 (1), 12−26.

Tellez, R., Gómez-Víquez, L., Meneses, A., 2012a. GABA, glutamate, dopamine and serotonin transporters expression on memory formation and amnesia. Neurobiol. Learn. Mem. 97 (2), 189−201.

Tellez, R., Gómez-Viquez, L., Liy-Salmeron, G., Meneses, A., 2012b. GABA, glutamate, dopamine and serotonin transporters expression on forgetting. Neurobiol. Learn. Mem. 98 (1), 66−77.

Terry Jr, A.V., Buccafusco, J.J., Wilson, C., 2008. Cognitive dysfunction in neuropsychiatric disorders: selected serotonin receptor subtypes as therapeutic targets. Behav. Brain Res. 195, 30−38.

Terry Jr, A.V., Callahan, P.M., Hall, B., Webster, S.J., 2011. Alzheimer's disease and age-related memory decline (preclinical). Pharmacol. Biochem. Behav. 99 (2), 190−210.

Thomas, D.R., 2006. 5-ht5A receptors as a therapeutic target. Pharmacol. Ther. 111 (3), 707−714.

Thomas, D.R., Soffin, E.M., Roberts, C., Kew, J.N., de la Flor, R.M., Dawson, L.A., et al., 2006. SB-699551-A (3-cyclopentyl-N-[2-(dimethylamino)ethyl]-N-[(4'-{[(2-phenylethyl)amino] methyl}-4-biphenylyl)methyl] propanamide dihydrochloride), a novel 5-ht$_{5A}$ receptor-selective antagonist, enhances 5-HT neuronal function: evidence for an autoreceptor role for the 5-ht$_{5A}$ receptor in guinea pig brain. Neuropharmacology 51 (3), 566−577.

Thomas, K.L., Everitt, B.J., 2001. Limbic-cortical-ventral striatal activation during retrieval of a discrete cocaine-associated stimulus: a cellular imaging study with γ protein kinase C expression. J. Neurosci. 21, 2526−2535.

Thompson, A.J., 2013. Recent developments in 5-HT$_3$ receptor pharmacology. Trends Pharmacol. Sci. 34 (2), 100−109.

Timotijević, I., Stanković, Ž., Todorović, M., Marković, S.Z., Kastratović, D.A., 2012. Serotonergic organization of the central nervous system. Psychiatr. Danub. Suppl. 3, S326−S330.

Tomie, A., Di Poce, J., Aguado, A., Janes, A., Benjamín, D., Pohorecky, L., 2003. Effects of autoshaping procedures on ^{3}H-8-OH-DPAT-labeled 5-HT$_{1a}$ binding and ^{125}I-LSD-labeled 5-HT$_{2a}$ binding in the rat brain. Brain Res. 975, 167−178.

Tomie, A., Tirado, A.D., Yu, L., Pohorecky, L.A., 2004. Pavlovian autoshaping procedures increase plasma corticosterone and levels of norepinephrine and serotonin in prefrontal cortex in rats. Behav. Brain Res. 53, 97−105.

Tsetlin, V., Utkin, Y., Kasheverov, I., 2009. Polypeptide and peptide toxins, magnifying lenses for binding sites in nicotinic acetylcholine receptors. Biochem. Pharmacol. 78 (7), 720−731.

Tsetsenis, T., Ma, X.H., Lo Iacono, L., Beck, S.G., Gross, C., 2007. Suppression of conditioning to ambiguous cues by pharmacogenetic inhibition of the dentate gyrus. Nat. Neurosci. 10 (7), 896−902.

Tsuruoka, N., Beppu, Y., Koda, H., Doe, N., Watanabe, H., Abe, K., 2012. A DKP cyclo (L-Phe-L-Phe) found in chicken essence is a dual inhibitor of the serotonin transporter and acetylcholinesterase. PLoS One 7 (11), e50824.

Upton, N., Chuang, T.T., Hunter, A.J., Virley, D.J., 2008. 5-HT$_6$ receptor antagonists as novel cognitive enhancing agents for Alzheimer's disease. Neurotherapeutics 5, 458−469.

van Praag, H.M., 2008. The cognitive paradox in posttraumatic stress disorder: a hypothesis. Prog. Neuropsychopharmacol. Biol. Psychiatry 28 (6), 923−935.

Vanover, K.E., Barrett, J.E., 1998. An automated learning and memory model in mice: pharmacological and behavioral evaluation of an autoshaped response. Behav. Pharmacol. 9 (3), 273−283.

Vanover, K.E., Harvey, S.C., Son, T., Bradley, S.R., Kold, H., Makhay, M., et al., 2004. Pharmacological characterization of AC-90179 [2-(4-methoxyphenyl)-N-(4-methyl-benzyl)-N-

(1-methyl-piperidin-4-yl)-acetamide hydrochloride]: a selective serotonin 2A receptor inverse agonist. J. Pharmacol. Exp. Ther. 310 (3), 943–951.

Volk, B., Nagy, B.J., Vas, S., Kostyalik, D., Simig, G., Bagdy, G., 2010. Medicinal chemistry of 5-HT5A receptor ligands: a receptor subtype with unique therapeutical potential. Curr. Top. Med. Chem. 10 (5), 554–578.

Wada, A., 2009. Lithium and neuropsychiatric therapeutics: neuroplasticity via glycogen synthase kinase-3beta, beta-catenin, and neurotrophin cascades. J. Pharmacol. Sci. 110 (1), 14–28.

Wallace, T.L., Ballard, T.M., Pouzet, B., Riedel, W.J., Wettstein, J.G., 2011. Drug targets for cognitive enhancement in neuropsychiatric disorders. Pharmacol. Biochem. Behav. 99 (2), 130–134.

Walstaba, J., Rappolda, G., Niesler, B., 2010. 5-HT₃ receptors: role in disease and target of drugs. Pharmacol. Ther. 128, 146–169.

Ward, B.O., Wilkinson, L.S., Robbins, T.W., Everitt, B.J., 1999. Forebrain serotonin depletion facilitates the acquisition and performance of a conditional visual discrimination task in rats. Behav. Brain Res. 100 (1-2), 51–65.

Wedzony, K., Maćkowiak, M., Zajaczkowski, W., Fijaø, K., Chocyk, A., Czyrak, A., 2000. WAY 100135, an antagonist of 5-HT₁A serotonin receptors, attenuates psychotomimetic effects of MK-801. Neuropsychopharmacology 23 (5), 547–559.

Williams, G.V., Rao, S.G., Goldman-Rakic, P.S., 2002. The physiological role of 5-HT2A receptors in working memory. J. Neurosci. 22 (7), 2843–2854.

Wilson, C., Terry, A.V., 2009. Enhancing cognition in neurological disorders: potential usefulness of 5-HT₆ antagonists. Drugs Future 34, 969–975.

Witty, D., Ahmed, M., Chuang, T.T., 2009. Advances in the design of 5-HT₆ receptor ligands with therapeutical potential. Progress Med. Chem. 48, 163–225.

Woehrle, N.S., Klenotich, S.J., Jamnia, N., Ho, E.V., Dulawa, S.C., 2013. Effects of chronic fluoxetine treatment on serotonin 1B receptor-induced deficits in delayed alternation. Psychopharmacology (Berlin) 227 (3), 545–551.

Woods, S., Clarke, N., Layfield, R., Fone, K., 2012. 5-HT₆ receptor agonists and antagonists enhance learning and memory in a conditioned emotion response paradigm by modulation of cholinergic and glutamatergic mechanisms. Br. J. Pharmacol. 167 (2), 436–449.

Woolley, M.L., Marsden, C.A., Sleight, A.J., Fone, K.C., 2003. Reversal of a cholinergic-induced deficit in a rodent model of recognition memory by the selective 5-HT₆ receptor antagonist, Ro 04-6790. Psychopharmacology 170, 358–367.

Xu, Y., Yan, J., Zhou, P., Li, J., Gao, H., Xia, Y., et al., 2012. Neurotransmitter receptors and cognitive dysfunction in Alzheimer's disease and Parkinson's disease. Prog. Neurobiol. 97 (1), 1–13.

Youn, J., Misane, I., Eriksson, T.M., Millan, M.J., Ogren, S.O., Verhage, M., et al., 2009. Bidirectional modulation of classical fear conditioning in mice by 5-HT₁A receptor ligands with contrasting intrinsic activities. Neuropharmacology 57 (5–6), 567–576.

Yun, H.M., Rhim, H., 2011a. 5-HT₆ receptor ligands, EMD386088 and SB258585, differentially regulate 5-HT₆ receptor-independent events. Toxicol. In Vitro 25 (8), 2035–2040.

Yun, H.M., Rhim, H., 2011b. The serotonin-6 receptor as a novel therapeutic target. Exp. Neurobiol. 20 (4), 159–168.

Zhang, L.N., Su, S.W., Guo, F., Guo, H.C., Shi, X.L., Li, W.Y., et al., 2012. Serotonin-mediated modulation of Na^+/K^+ pump current in rat hippocampal CA1 pyramidal neurons. BMC Neurosci. 13, 10.

Zhou, W., Chen, L., Paul, J., Yang, S., Li, F., Sampson, K., et al., 2012. The effects of glycogen synthase kinase-3beta in serotonin neurons. PLoS One 7 (8), e43262.

Zhu, B., Chen, C., Loftus, E.F., Moyzis, R.K., Dong, Q., Lin, C., 2013. True but not false memories are associated with the HTR2A geneNeurobiol. Learn. Mem pii:S1074-7427(13)00181-0. Available from: http://dx.doi.org/10.1016/j.nlm.2013.09.004. [Epub ahead of print] PMID: 24055687.

Zimmer, L., Le Bars, D., 2013. Current status of positron emission tomography radiotracers for serotonin receptors in humans. J. Label Compd. Radiopharm. 56, 105–113.

Zovkic, I.B., Guzman-Karlsson, M.C., Sweatt, J.D., 2013. Epigenetic regulation of memory formation and maintenance. Learn. Mem. 20 (2), 61–74.